Graham Nerlich

Einstein's Genie
Spacetime out of the Bottle

MINKOWSKI
Institute Press

Graham Nerlich
University of Adelaide
graham.nerlich@adelaide.edu.au

Cover: Bottle without genie. Designed by David Nerlich.

© Graham Nerlich 2013
All rights reserved. Published 2013

ISBN: 978-1-927763-13-1 (softcover)
ISBN: 978-1-927763-14-8 (ebook)

Minkowski Institute Press
Montreal, Quebec, Canada
http://minkowskiinstitute.org/mip/

For information on all Minkowski Institute Press publications visit our website at http://minkowskiinstitute.org/mip/books/

For Angas Hurst AM FAA
22nd September 1923 to 19th October 2011
Physicist, philosopher, friend.

If Aladdin called forth from the bottle a genie who was unwilling to return, Einstein gave to geometry a life and an independent existence, which he was equally astonished to see take shape before his eyes.

J. A. Wheeler in the Foreword to Graves (1971:vii).

Preface

This book is about the Theories of Relativity, both Special and General. What intrigues me about them is the role of spacetime – intriguing because many doubt that spacetime can play any role. They regard spacetime as unthinkable – metaphysically impossible. Yet the orthodox view of the General Theory is that it dispenses with the force of gravity by reducing it to the curvature of spacetime. Sections §4 to §22 of the first epoch-making paper of 1916 amply justify that view. Yet Einstein himself blinked. The first three sections of that first paper show that he did not understand the theory as delivering that result but rather of ridding the theory of the metaphysical baggage of space and time. By the next year the mathematicians showed that this claimed too much. Despite the triumphal expulsion of the classical ether from Special Theory, Einstein argued in later work for replacing spacetime with an ether; later still that spacetime is really the gravitational field.

My belief is that Einstein recoiled not because the General Theory really is unthinkable but because he could not countenance the metaphysics that the direct interpretation of the 1916 paper's mathematics and physics commits one to. His hostility to spatial ideas pervades much his thought and it was based partly on his conviction that space, time and spacetime can't be observed and partly on the thought that space is somehow or other just nothing. The task of this book is to show that spacetime is thinkable just by rethinking it through with that aim in mind.

Einstein insisted that scientific theory was a free creation of the human mind and believed that his own most significant talent was his creative imagination or fancy. That talent was exercised in a way never surpassed in power, scope and boldness both in his 1905 paper on Special Theory and in his 1916 paper already mentioned. If I succeed in showing that spacetime is thinkable after all then we must conclude that Einstein's doubts were a failure of metaphysical imagination.

So the book is a philosophy book, despite all the discussion in it about relativity physics. Its main ambition is to show something about the power of the human mind through a critique of possibly the finest ever exercise of it. I do not offer a theory of imagination in general. That is beyond my imaginative powers. But to defend spacetime as a splendid instance of rational imagination is my basic aim.

That's a stark and crude picture of my goal and it takes a whole book to give something like an adequate account of it. Most of the book explains how spacetime does work in relativity theory and much of that explanation can be given just by looking closely at the simpler Special Theory. Einstein's doubts, and the doubts of other later physicists and philosophers about the General Theory are not put under the microscope until chapters 9 and 10 after the positive account of spacetime's role has been given and explained. By then the metaphysical source of Einstein's concerns can be laid bare and one can estimate the strengths and weaknesses of these objections to the orthodox view.

I hope that those broadly interested in the physics of relativity will find the book stimulating although there is no new physics in it. I have completely ignored quantum gravity since that obscures a clear picture of what pure General Relativity says about spacetime. That enables me to include non-specialist readers who already have at least some insight into the kind of theory relativity is, as well as physicists proper. I aim to engage both philosophers of science and philosophers with a more general metaphysical bent. There are almost no equations and only a few formal expressions throughout. This is not a popular exposition of the theory but I have been at considerable pains to make the book accessible to any readers who would be seriously interested in it. There is a good deal of exposition both to aid readers and to aid critics by showing how I think the fundamental ideas work. Parts of chapters 9 and 10 are written at a more advanced level than the rest of the book but, by then, almost all the positive ideas have been explained and defended.

I wrote this book in retirement for the pure strenuous fun of it. I hope the product reflects that. It would have been much less strenuous and much less fun without weekly meetings over 18 years with the circle of friends, philosophers and physicists that is the Fizz group. I owe much and am deeply grateful to: Angas Hurst (now dead alas), Peter Lavskis, Steve Leishman, Colin Mitchell, Chris Mortensen, Peter Quigley, Michael Simpson and Peter Szekeres.

To other friends, for support, discussion and correspondence, I am grateful to Rick Arthur, Barry Dainton, Brian Ellis, John Lucas, Steve Lyle, Hugh Mellor, Greg O'Hair, Vesselin Petkov, Oliver Pooley, Huw Price and Jack Smart.

The figures and the cover design of the Klein bottle in an Arabian desert landscape were contributed by my son David Nerlich. I warmly thank him for them.

For her encouragement, patience, love, good humour and a remarkable ability to keep me happy I thank my wife Margaret Rawlinson. Without whom, nothing. I thank the Minkowski Institute for doing me the honour of including me among its Founders.

I find myself in frequent and fundamental disagreement with those from whose work I have learned most about relativity and its philosophy. Of course Einstein and Minkowski are first but not alone among those to whom I owe so much. I hold them all in deep respect and admiration despite differences. I have intended nothing other than friendly, respectful, if sometimes unequivocal, dissent.

I am grateful to the Australia Research Council for their support in funding this research.

Chapters 7 and 8 are revised versions of earlier papers with the same titles. "How the Twins Do It: Special Relativity and the Clock Paradox" *Analysis* **64**, (2004): 219-230.

"Why spacetime is not a hidden cause: a realist story" in V. Petkov (ed.) *Space, Time and Spacetime Physical and Philosophical Implications of Minkowski's Unification of Space and Time*: Springer: 177-188.

Adelaide, July 2013   *Graham Nerlich*

Contents

0 Introduction   1
- 0.1 What this book is about 1
- 0.2 Einstein's reputation and spacetime's part in it 3
- 0.3 The dark side of the genie 5
- 0.4 Ontology and ontic type 6
- 0.5 Spacetime and quantum theory 10

1 How the Leibniz Shifts Backfire   11
- 1.1 The dazzle of Leibniz shifts 11
- 1.2 Symmetry or identity? 13
- 1.3 Shifts in spherical space: S_2 and S_3 14
- 1.4 Other shifts 16
- 1.5 Detachment and a dualism among relations? 18
- 1.6 On the importance of regions 18
- 1.7 The unnatural longevity of the shift arguments 19
- 1.8 Who may exploit the shifts? 20
- 1.9 Homogeneous spaces and shift symmetries 22

2 Path Realism: The Traditional Debate Reshaped   27
- 2.1 The traditional debate 27
- 2.2 What makes spatial relations spatial? 28
- 2.3 An entrenched belief 31
- 2.4 On primitives 32
- 2.5 Paths and perception 33
- 2.6 Distance without paths 36
- 2.7 Taking "intrinsic" seriously 38
- 2.8 Substrata 40
- 2.9 Waddaya mean – concrete-immaterial? 42
- 2.10 The modesty of realism 43
- 2.11 Reflexive spatial relations 45
- 2.12 Talking about paths 45

3 On the Sovereign Independence of Spacetime   47
  3.1 Introduction 47
  3.2 Minkowski's "purely mathematical line of thought" . . . 49
  3.3 The form of a new metric 50
  3.4 The value of c 53
  3.5 A deductive approach 55
  3.6 Field theories and SR 56
  3.7 Electromagnetism 57
  3.8 Mechanics 58
  3.9 Metaphysics, not physics 58

4 On the Benefits of Four-Thought   61
  4.1 Introduction 61
  4.2 Representing worldlines 61
  4.3 Representing mass, momentum
        and energy: $e = mc^2$ 63
  4.4 Speeds, angles, rapidities 65
  4.5 Mass and energy in systems of
        particles 67
  4.6 Conservation in collisions 68
  4.7 Radiation 72

5 In SR, Why Does A Moving Rod Contract?   75
  5.1 Introduction 75
  5.2 On constructive explanation 76
  5.3 Bell's Lorentzian Pedagogy 78
  5.4 What Bell said 80
  5.5 A conjectured interpretation 81
  5.6 The question of acceleration 83
  5.7 What explanation lies at the end of the "long road"? . . 84
  5.8 The puzzle of the thread 86
  5.9 Difference of style 88
  5.10 What is a relativistic rod? 89
  5.11 What is SR? 90
  5.12 The dynamical interpretation of
        contraction 92
  5.13 Lorentz contraction and spacetime 94

6 Time and Spacetime: the Same Ontic Type   99
  6.1 Introduction 99
  6.2 Time diagrams 100
  6.3 Temporal solipsisms 104
  6.4 Existence, occurrence, truth, reality 107
  6.5 A naïve model of naïve time 110
  6.6 Time, space and spacetime: the same in ontic type . . . 112

7 How the Twins do it: SR and the Clock Paradox   113
    7.1 Introduction 113
    7.2 What is the twins paradox? 114
    7.3 3D description and the A-theory 115
    7.4 A 3D B theory explanation of the paradox 116
    7.5 Frames of reference 118

8 Why Spacetime is not a Hidden Cause: a Realist Story 123
    8.1 Introduction 123
    8.2 Cause and classical inertial motion 124
    8.3 The law of motion 127
    8.4 Space, time and spacetime in GR 128
    8.5 Free fall in a purely gravitational field 130
    8.6 Light bending 132
    8.7 About matter 133
    8.8 A parody of "hidden cause" 134
    8.9 Does matter act on spacetime, telling it how to curve? . 137

9 Is Spacetime Really Spacetime?   139
    9.1 Introduction 139
    9.2 The ether: Einstein and others 141
        9.2.1 The concept of field 141
        9.2.2 Principle of Equivalence 142
        9.2.3 Action at a distance and the ether 144
        9.2.4 Two modern views 146
    9.3 The fundamental equation 148
        9.3.1 The Einstein tensor G 149
        9.3.2 The matter tensor $T_{\mu\nu}$ 149
        9.3.3 Pre-eminence of the metric 151
    9.4 Seeds of doubt – energy without work? 152
        9.4.1 GR is a force-reduction theory 153
        9.4.2 Energy, force and work 154
        9.4.3 Conservation failure and forms of energy in curved space and spacetime 155
        9.4.4 Hard cases 155
    9.5 Total, gravitational and T-sourced energy 158
        9.5.1 Regional energy in curved spacetime 158
        9.5.2 Imposing time-translation symmetry 159
        9.5.3 Dark energy 160
    9.6 Conservation problems 160

10 The Trouble with General Covariance — **165**
- 10.1 Introduction 165
- 10.2 General covariance: an introduction 166
  - 10.2.1 The case of the Euclidean plane 166
  - 10.2.2 The case of GR 170
- 10.3 Theories, models and worlds 172
- 10.4 Approaching the Hole Argument 174
  - 10.4.1 Active and passive transformations 174
  - 10.4.2 Leibniz shifts vs diffeomorphisms 175
  - 10.4.3 On formulation and determinism 176
- 10.5 The Hole Argument 177
- 10.6 Critique of the Hole Argument 178
  - 10.6.1 The models really are distinct 179
  - 10.6.2 A serious embarrassment? 180
  - 10.6.3 A philosophical issue? 181
- 10.7 Einstein on general covariance, 1915 and later 184
- 10.8 The role of the manifold 187

11 Concrete Yet Insubstantial — **189**
- 11.1 Introduction 189
- 11.2 Cause and causal powers 190
- 11.3 Sophisticated substantivalism
  and structuralism 193
- 11.4 Points and parts of spacetime 193
- 11.5 Concluding remarks 195

References — **197**

Index — **209**

About the author — **215**

0 Introduction

0.1 What this book is about

This is a book about spacetime. If Einstein's relativity theories are true then spacetime is something real. But what kind of real thing is it? I aim to explain and justify various aspects of the problem that lead to a definite answer. It is easy enough to say what kind of thing spacetime is not. It's the positive story that challenges. Obviously neither space, time nor spacetime is a hardware thing like a lump of matter or any kind of a liquid or a gas. Less obviously and much more contentiously, it is not a system of spatial relations just among observable hardware things. It isn't an abstract mathematical entity like a number or a set of them. It isn't an ether of any of the kinds that crop up in the literature early and late. It isn't like minds or numbers or gods. In fact it is no kind of substance or stuff at all. To do the job needed in General Relativity it need not, indeed cannot be substantial.

Space, time and spacetime, if they are real, are rather like each other but utterly unlike anything else. They belong to their own unique type. That negative list partly explains why the positive story always gets a bad press, since the list seems to exhaust the kinds of thing there are or could be. Clearly then, this main question – what is spacetime? – is a metaphysical one belonging specifically in ontology. We may rephrase our question as "What is the *ontic type* of space, time and spacetime?" That assumes that they are usefully seen as of the same type. A bit more detail on the loose, but useful idea of ontic type comes later in this Introduction (in §4).

Spacetime is crucially important to metaphysics. It plays a strong, if puzzling, role in ontology. There is a long-lived, widespread but contested belief [i] that everything real is contained in spacetime – all hardware things, as well as liquids and gases – but it is not contained in anything else [ii] anything that purports to lie "beyond" it must fail to *make sense* as a real thing. E.g. God or numbers make no good sense. Yet the existence and nature of spacetime itself are highly contentious. It constitutes a boundary for what's real and cannot, itself, fall on either side of it. I don't take for granted these beliefs and that picture of spacetime's role in ontology. The issues need deeper consideration than I shall give them, but they serve to heighten from the start a sense that spacetime is centrally important in metaphysics.

The right place to look for an answer to our question is in the theories of relativity, Special and General (hereafter SR and GR respectively). GR is hard physics and hard maths. But I pursue the theories only as far as they tell us what spacetime is. We dig only round their foundations, a fascinating exercise but one that calls for no skills in solving equations, although some understanding of mathematical and physics ideas. I write for philosophers generally (and for others who find the question teasing) not only for philosophers of science. But this is an essay in philosophy of physics.

The relativity theories provide an interesting and quite revolutionary answer: spacetime is an entity that explains what gravity is by reducing it to spacetime's curvature: thereby spacetime geometry shows how classical *dynamical* explanations of gravitation give place to *kinematical* ones. This can be made clear in the end provided only that we allow spacetime to be a thing of an ontologically quite unique kind. But some of the influential literature relevant to the question shows something else. From the start a tradition began of claiming that GR shows that spacetime may be dispensed with and that the theory is really about something else – most often that spacetime is really gravity or an ether! There are disputes about the direction and kind of explanation in the theories: they need investigation. In his very first paper on GR in 1916 Einstein tells us that the theory "takes away from space and time the last remnant of physical objectivity" (Einstein, 1916:117). So, if Wheeler's analogy between Einstein and Aladdin is apt, it is also the case that Einstein was eager to get the genie of spacetime safely back in the bottle. What is really a profoundly original theory of geometro*kinematics* became something that was called, for a while, geometro*dynamics*. If gravity does turn out to be a force then it will be a strange one: it is absurdly weaker than any force (force field) recognised in the quantum theory Standard Model's array of forces and fields. It accelerates all objects in the same way irrespective of their constitutions or their mass; it has no anti-force; there are no insulators against it.

My claim throughout is that GR is indeed a theory in which spacetime is the main player and to understand it is to understand its explanatory role. It does reduce gravity to the variable curvature of spacetime. How it does that and how the explanation works is a central problem. *Gravitation* (Misner et al., 1973) says throughout that gravity reduces to spacetime structure but when it comes to saying how the explanation works there are prominent obscurities: "Spacetime acts on matter telling it how to move. Matter acts on spacetime telling it how to curve". But spacetime cannot act or tell and it is neither acted upon nor told anything. These fog patches arise from confusion as to the kind of thing spacetime is. Free fall has to be explained as free – free of forces, of actions, of telling and of being told. But matter in a gravitational field is never really free and it never simply follows geodesics. *Gravitation* focuses both sharply and confusedly on these matters. It is no part of GR to reintroduce a gravity field as a dynamical explainer. I don't say that cannot be done, but it is no longer really GR when it is done.

Less contentiously, GR may be understood as a dynamical or as a kinematical theory. The philosophical interest is that it can be understood in the latter way as a theory in which gravity is not a force but a structure of spacetime. On that understanding it is much the more revolutionary, elegant and interesting form of GR.

I will argue that the trouble at this foundational level lies in spacetime's ontic uniqueness. The same is true of space and of time. That it is not substantial yet it is concrete sounds like a metaphysical paradox. My task is to show how those claims make sense together, that failure to grasp how they do leads to obscurities and how admitting it solves problems. It is no light task. My final aim is to give an extended, clear and convincing account of it.

The truth of GR is not my guiding concern. I aim to find the metaphysical type of spacetime as GR portrays it. That picture revolutionised our grasp of the kinds of entity that could go to make up a realistic world-picture. So the main problem is to grasp how something could earn the central place in one of our two best theories of the universe while being utterly unlike the things philosophy easily countenances.

0.2 Einstein's reputation and spacetime's part in it

Albert Einstein appears on the cover of Time Magazine's first issue in 2000 as "Person of the Century". We may disagree with that choice, but surely it doesn't surprise us. He towers over the science of the 20th century. In that century "Einstein" was everywhere a household word. It still is.

Within the decade 1905-15 Albert Einstein first published the Special Theory of Relativity (Einstein, 1905). There followed a long struggle to construct General Relativity, the most powerful and beautiful theory ever to grace science, with its completion in 1915 and publication in 1916. The two relativity theories are fundamental, global in scope, elegant in conception – and astonishing. These attributes place them among the great cultural achievements, possibly the greatest, in human history.

Yet it is not quite obvious why Einstein was Person of the 20th century. I have no wish to challenge his entitlement but rather to endorse it. I ask only what grounded it.

The consequences of the relativity theories impinge negligibly on our daily lives compared with the pervasive influence of electromagnetism. Electromagnetic gadgetry of more or less complex kinds surrounds us every hour, every minute. More than any change in history, what Faraday and Maxwell discovered in the 19th century revolutionised daily practical life, both in the kitchen, across the globe and beyond it to observation of the stars.[1] Yet those who know of Faraday and Maxwell are rather educated folk. Einstein is far more prominent as a revolutionary. But his science is remote from everyday affairs. The working of GPS systems is one of the very few, if not the only, application of GR to practical everyday problems.

SR began with queries about the coherence of electromagnetism as it was understood before 1905. GR tackled the formidable problem how to find a theory of gravity consistent with it. Its payload was theoretical not observational and it cited no current work. For decades GR explained only the observed, tiny, but significant discrepancy in the classical prediction of the advance of the perihelion of Mercury. The only new observations predicted by the theory were the bending of light in a solar eclipse, the gravitational red shift in radiation and the later observed expansion of the universe.

Relativity physics does not shock our view of our status in the universe as the 16th and17th century's displacement of man from the centre of the universe did. It doesn't match the 19th century revelation that species have evolved so radically that man has inhuman ancestors. These earlier revelations admit reverberating epigrams: we live on the modest planet of an insignificant star; we are a rather smart kind of ape. Relativity says nothing to match that.

I suggest that Einstein's stellar reputation stems from the central role of spacetime – immaterial, kinematic, concrete, non-dynamic. Einstein's international fame began with Eddington's measurement of the bending of light rays in solar eclipse and the statement that this sprang from a curvature of spacetime. Einstein dared to think that, was able to invent and understand it and see its consequences.[2] Both for the learned and for the folk, the wonder of Einstein's creative imagination is the wonder of spacetime and its consequences. Space, and

---

[1] It is said that a minister in government, who visited Faraday's laboratory, asked, "But Mr. Faraday, what is the use of it?" Faraday replied, very pertinently given the trade of his inquisitor, "Well, sir, you will be able to put a tax on it."

[2] Wheeler saw this as the mainspring of Einstein's fame (see (Wheeler, 1998:230-1).

more especially time, have always mystified us. Even the simple idea of a four dimensional world still induces in many a certain awe and bafflement. Still more do the following: the Clock (or Twins) Paradox, the contraction of a moving rod, the curvature of spacetime in all its dimensions, that time runs slower at the surface of the earth than in empty space and slower still at the surface of the sun: Later, in addition to all these the famous equation $e = mc^2$ seems to inhabit an unbreathable air of intellectual altitude. These all spring from the union of space and time in spacetime and its role in GR. That is the core of both the popular and the learned sense of his genius.

If we construe GR as without its original and simplest commitment to spacetime then it becomes a theory of gravitation as a force that is a more complex and accurate version of pre-1916 theory. The 1905 SR is wonderfully imaginative but without the conceptual novelty of spacetime as a major player one might think Maxwell's electromagnetism a more daring and innovative theory than GR.

So the hero of this book is impersonal – spacetime. It was only implicit in Einstein's SR. But Hermann Minkowski (Minkowski, 1908) saw clearly just what was encapsulated. He was first to realise that spacetime transcended the theory of light and electromagnetism, resting on properties belonging to space and time themselves – simple properties, familiar from Euclidean geometry and the assumption that the laws of physics don't change; i.e. they are time translation invariant. (This is the topic of Chapter 3.) Minkowski provided a quite different foundation for SR, a fact largely ignored since. However, it was Einstein who then pulled the cork in relativity physics, released the genie, spacetime, and put it to work as the hero of GR. Spacetime's role was dominant, unprecedented and remains unique in its distinctive, geometric style of explanation. Briefly no one before had seriously tried to use spatiotemporal concepts to replace core ideas of dynamics with simpler ones of pure kinematics.[3] It gave the theory not just depth and power but also novelty, astonishment, beauty and cosmic scope. And a touch of magic.

0.3 The dark side of the genie

Einstein valued above all else the free creativity of intellect – flights of disciplined imagination.

> When I examine myself and my methods of thought, I come to the conclusion that the gift of fantasy has meant more to me than my talent for absorbing positive knowledge.

---

[3] W. K. Clifford (1995) had attempted to exploit the geometry just of space. It was an unworkable but not entirely dissimilar ambition.

Cited in Taylor and Wheeler (2000:B-10)

> Science is not just a collection of laws, a catalogue of unrelated facts. It is a creation of the human mind, with its freely invented ideas and concepts.

Einstein and Infeld (1961:294)

His soaring imagination freely created spacetime as the main and conceptually. the newest and most daring, player in GR. It really was like putting a genie to work! It is an unfortunate paradox that he persistently wished to repudiate what he released. I will argue that this was a failure of metaphysical imagination.

Philosophy should be a custodian of reason as well as truth: we have a core interest in how imagination may cross the bounds of sense and stray into nonsense. So we need to explore which constraints on invention are proper. The problem troubled Einstein, the most philosophical of physicists.[4] His admiration for Mach's critique of Newton's absolute space and time convinced him that space and time have no place in objective physics. There are only relations among material things. Being beyond observation, spacetime is beyond sense. Crudely Einstein always thought that space and spacetime were unthinkable. The task of this book is to show that he did indeed think them and that he underestimated the powers of human thought even when he exercised them superlatively well. My aim is to explain this and make some comment on human intellectual imagination through exploring this example of it.

0.4 Ontology and ontic type

What worried Einstein, and it has worried many another since, was something metaphysical: the disconcerting *ontic type* of spacetime. What kind of thing could it possibly be? That's an ontological question since it's about what broadest types of things there are. It also raises an epistemological question: whether and how we can know about them. That issue is somewhat tangential to the concerns of this book.

Metaphysics is and always was a chaotic battleground:

> And we are here as on a darkling plain
> Swept with confused alarms of struggle and flight
> Where ignorant armies clash by night.

M. Arnold "Dover Beach"

---

[4] Among scientists only Einstein has a volume in P. A. Schilpp's series *Library of Living Philosophers*. See Schilpp (1949).

Perhaps it's not that bad but attempts to make it orderly seldom enjoy long lives. Nevertheless it will help to sketch one scheme for ontology, the theory of ontic types.[5]

Descartes proposed an ontology with just two ontic types, mind and body. It shaped the whole of his philosophy. He saw the distinction as exclusive and exhaustive: everything was a mind or a body; nothing was both. The peculiar, salient and most general property of mind was to think, of body to be extended. There was no property, more general than these two, that both could share. This illustrates the kind of scheme an ontology might have - an exhaustive list of exclusive ontic types with different types utterly unlike each other. That lets me put my thesis in stark if crude terms: spacetime is unique in ontic type, a type exclusive of everything (save its aspects, space and time) and fills a salient place in any list of types that purport to exhaust the things that are real. The idea of ontic type is not asked to carry much load but only to sketch the kind of thesis I defend.

A bad odour from a history of extravagant, obscure or bogus ontic types demands that each candidate be examined for good sense. Some types – minds for instance – that are undeniably real, seem to be of their own type because they have unique and puzzling general properties. They may continue to worry us even when we place them in some familiar basket (e.g. in with the brain), or deem them a type of which they are the only member (as Descartes proposed). Candidate ontic types include minds, gods, numbers, moral values, as well as spacetime.

Materialism decisively failed with the advent of electromagnetism giving place to the idea that there are matter *fields*. The concept of field plays a prominent role in modern physics. A field in space or spacetime is a quantity attached to each point. The quantity may be simple or complex: there are scalar fields, vector fields and tensor fields. Usually the interesting field quantity is dynamical – a force field – but that is limiting as a defining property. The quantity may be said to be physical but this raises difficulties as to what counts as physical or material. I mention these caveats early since the geometry of space or spacetime is itself a field – the metric field. Whether that field is physical just because physics books talk about it or whether it is from being connected more intimately with some concept of matter or dynamics is both a subtle question and a contentious one. (See Chapter 9.)

---

[5]DiSalle (2006) approaches the ontology of spacetime through the dialectical history of the concepts of space and time up to and including spacetime. This limpid and penetrating book aims to avoid the traditional debate between substantivalism and realism whereas I try to pass between the horns of that dilemma and resolve it. The end results are somewhat different. Without meaning to compare my *success* in the one approach with his in the other I see the *approaches* as consistent and equally valuable.

Epistemology studies how we know – if we do – about the different ontic types of things. Quine (1960) advised that we should admit to our ontology those entities that our best theories of the world oblige us to admit. But this good advice takes no heed of the possibility that a good theory might, nevertheless, oblige us to admit things whose ontic type is worrying. That is how things stand with spacetime in GR, despite its prominent role in that theory. The theory is excellent. It is the nature of spacetime that gives us pause. The main worry is that space, time and spacetime are said to be unobservable: that is addressed in detail later. Quine's counsel is small help with spooky ontic types. We've all got minds according to any decent theory. But that helps little in puzzling out what kind of thing a mind is! Further it is not quite clear which our best theory of mind is or even whether the seemingly relevant ones are much good. Our problem with the genie is that while GR is outstandingly good, and while it freely deals with spacetime,[6] it is hotly disputed by many what spacetime is and therefore whether the theory really demands it.

What is the problem with the ontic type of spacetime? Mainly it is hard to see how it can be anything at all rather than nothing. Space used to be called the Void, mere emptiness, no thing. Yet it's not a straightforward fiction, like the usual run of nonentities: witches, ghosts, or alien abductors. All of us talk seriously about times and places, distances and periods. Scientists freely discuss intervals, trajectories and so on: the differential calculus is heavily dependent on spatiotemporal concepts. Unlike the theoretical concepts of quantum physics, space and time are familiar. Spacetime needs to be explained: but like its forebears, it pervades perception; it is not too small, nor too remote; it's not undetectably short-lived, it doesn't act instantaneously across a distance. It is not hidden but, in a real sense, apparent. You don't see it but you do see across it: if you didn't you wouldn't see anything at all. You move, spin, somersault and so on, in it. It is never mistaken for anything else, as phlogiston was mistaken for oxygen. If it's anywhere it's everywhere; if at any time then at every time. It explains much and its explanations, unlike magical ones, do not depend on illusion or mystification. The range of spacetime's performance is cosmic. In GR, it fixes the motions of planets and stars; its geometry collapses into black holes yet also admits an expanding universe. Nothing else does anything like that. Yet, for many, real spacetime is just bad metaphysics. Its style of explanation is not always thoroughly understood.

Does time belong to the same ontic type as space and spacetime? They would be strikingly different if time really passed or flowed as we

---

[6]Misner et al. (1973) write of spacetime in a very direct and committal way throughout. It is what that massive and authoritative book is about.

commonly say it does. The idea of a flow or passage of time has strong intuitive appeal but it is a notoriously thorny problem to articulate what this could mean. The issue has been widely and intensively discussed without reaching a strong consensus. I endorse the B-theory of time. There are processes and events in time but *time* is not in time. Things endure and change in time but time itself doesn't change nor does it endure unchanged. Changes take time and it takes time to be static – to remain unchanging. There is no time outside time or spacetime in which time or spacetime could either change or be static. The B-theory should not admit that time is static for the same reason that it should not admit that time changes.

The key to ontic likeness is that there is no difference in ontic status among times just as there is none among places. In that last sentence "is" must be understood, not as a present tensed, but as a tense-free form of the verb "to be." To say that past events are real is not to say that they are real now but just that earlier events really happened. Grammar demands that our verbs be formally tensed but not that "is" means "is now."

The semantics of "now" is like that of "here." Both expressions are indexical: a rule of reference for "now" is that, on any occasion of utterance, it refers to the time of the utterance.

We often draw spatial diagrams of time in which spatial extension represents temporal extension. Minkowski diagrams do this. It is easy to feel their inadequacy. Whereas the spatial dimension is represented by a bit of space no diagram on a page can represent time as bit of time. Yet, inadequate though diagrams may be in this regard, they need not mislead us unless we feel that we can only repair the inadequacy by somehow adding a point or a line to represent a present and having it move up the page in real time. That invites confusion since time is represented in two quite different ways, once spatially and once temporally. To sum up this brief discussion in an epigram: space is that in which things are separated and connected with each other by spatial relations: time is that in which events are separated and connected with each other by temporal relations. To grant a putative entity a share of the ontic sunshine, you need to be clear and specific as to what it is. That is needed, too, to delineate how spacetime does its explanatory work. To describe it in sufficient detail is a main task in this book.

0.5 Spacetime and quantum theory

Relativity and quantum physics have remained unreconciled for over a century. One or both of them is in error but which one, and how remains the most urgent problem in physics. The smart money is that GR will change more than quantum physics will. So "quantum gravity" names the enterprise of reformulating GR so as to accommodate an operator formalism and quanta – this despite the fact that GR is by far the more intelligible and beautiful structure. This has resulted in so many attempts to reformulate GR that the first pure form of the theory is easily lost sight of in the welter of revisions, none of which has yet done the job of bringing the two theories into harmony. Some quite new form of GR will doubtless emerge under these strains, but, for now, spacetime remains the heart of a supremely imaginative construction of a human mind that has been crowned with massive successes. If its looming revisions make it history rather than truth, then, in the long run, we are all dead and everything is history. That holds for any scientific theory at any stage. But neither we, nor GR, are dead yet.

1 How the Leibniz Shifts Backfire

Chapter 1 is a critique of the most influential and persuasive style of argument against the reality of space, the Leibniz's shift objections. I mainly consider modern, pruned versions not those in Leibniz's famous correspondence with Samuel Clarke. I argue that these overlook the range of geometries possible for space. Once that is noticed, the shifts fail to sustain the conclusion even in the most favourable instance of Euclidean space. I draw the moral that one can't detach spatial relations among observable things from the geometry of the space they occur in. Appeal to shifts backfires on relationists since they make no sense save against the very thing they seek to be rid of – a space with a specific geometric structure.

1.1 The dazzle of Leibniz shifts

In 1715-16 Leibniz advanced his remarkable shift arguments in order "to confute the fancies" of those who take space to be a real thing. He wrote them in a famous correspondence (Alexander, 1956) with Samuel Clarke who was Newton's spokesperson. John Norton describes Leibniz as "taunting" Newton with these arresting objections to absolute space as Newton described it (Norton, 2011:§5). Neither Newton nor Clarke betrayed embarrassment in replying to them. Yet these objections have succeeded in setting the stage for discussion of the reality of space ever since their publication. They are the most elegant, most famous and most influential arguments in all the philosophy of space and spacetime. Leibniz aimed to expose the idea of a space over and above spatial relations of things to things as a nonsensical monstrosity – a chimera. Let us focus on just one of three arguments that he advanced, called the static shift. Here are two versions of it, the original and a modern one.

Space is something absolutely uniform; and, without the

things placed in it, one point of space does not absolutely differ in any respect from another point in space. Now from hence it follows, (supposing space to be something in itself, besides the order of bodies among themselves,) that 'tis impossible there should be a reason why God, **preserving the same situations of bodies among themselves**, should have placed them in space after one certain particular manner, and not otherwise; *why everything was not placed the quite contrary way, for instance, by changing East into West.*

(Alexander (1956:Fifth paper, §5); my italics describe the shifts; my bolds their outcome.)

Leibniz's phrasing, "by changing East into West," is ambiguous between reflecting and rotating the one into the other. Perhaps this is why, in a more recent version by Earman and Norton – the real target of this chapter – it seemed better to substitute a shift of everything the same distance in the same direction. This version occurs in their paper on the Hole Argument, a dominant theme in spacetime philosophy since 1987:

Whatever reformulation a substantivalism may adopt, they must all agree concerning an acid test of substantivalism, drawn from Leibniz. *If everything in the world were reflected East to West (or better, translated 3 feet East)*, **retaining all the relations between bodies**, would we have a different world? The substantivalist must answer yes since all the bodies of the world are now in different spatial locations, even though the relations between them are unchanged.

(Earman and Norton (1987:521); my italics describe the shifts; my bolds describe the claimed outcome.)

First, notice that, as these three authors frame it, the objection is against the general idea of space, not just of Euclidean space. In the context of 17th century thought, neither Leibniz nor Newton ever supposed, or could have supposed, that space might be other than Euclidean It was never thought of as a special case. Leibniz is explicit: "without the things placed in it, one point of space does not absolutely differ in any respect from another point in space. Now *from thence it follows* ..." (My emphasis). The premise states a property of *all* Riemannian spaces. The conclusion does not follow. In the context of the Hole Argument, the target of discussion is the topological basis of all the metric spaces admitted in GR. (See Chapter 9). No special case, such as Newton's absolute space, is intended or relevant for their purpose.

The authors assume that we all want to answer the acid test question "no; the worlds are not different." So the substantivalists must give the wrong answer – "yes." However they may wriggle, they can't get off the hook.

There are other strong and striking claims and arguments of very much the same kind. Briefly, if the velocity of everything in the universe were increased by the same amount, in the same direction, at the same time, that would result in exactly the same world. That is Leibniz's *kinematic* shift. Then if everything were accelerated in the same way, that, again, would still be the same world. That is the *dynamic* shift. Finally, there is a shift suggested by ideas of Poincaré rather than Leibniz. If everything doubled in size overnight, no discernible difference would result. It would still be the same world.

1.2 Symmetry or identity?

Without Leibniz's contentious appeal to God's reasons, how is the objection valid and how do identity and difference get into the argument? Earman and Norton ignore the theological premise. That is not the issue. The immediate upshot of a proposed shift is that it yields a symmetry: that is, a state of affairs that is exactly like the beginning state in some significant way. But to banish the relations of things to space – that is the aim – we need more than just the symmetry. Leaving aside the dubious Principle of Sufficient Reason, Leibniz himself had a definite proposal about this: The Principle of the Identity of Indiscernibles. It says that if two things have all their properties, including relational ones, in common they are identical and not two. So unless we can discern one thing from another by saying how they differ in properties or relations, we should conclude that they are one.

However, the symmetry by itself does not strictly yield the indiscernibility since each thing is in a *different* place. Before the argument, that made sense and the shift to a different position made a different world. Thus the states of affairs (or possible worlds) do differ – in the relations of things to space in each case. That is how substantivalists must answer and how Clarke did answer. Yet that answer may well make us uneasy since, inside the worlds, we could observe no difference. If we still want the identity conclusion, we might say that thing-to-space spatial relations somehow do not count as making differences for the purposes of the Principle. Whether or not that is true, we can't assume it here because just that is what the argument sets out to convince us of.

This raises a more general problem: what decides what will count in applying the Principle? How exactly should the Principle be formulated? How and why should observation play a role?

While these worries about the Principle are pertinent, should they sway us? Classically, reduction-to-absurdity arguments should conclude with a *contradiction* as the absurdity. These arguments do not. Nevertheless Leibniz's examples have proved immensely appealing. Their persuasive power, resting on the claim that we could not know that there is a difference, has proved strong and durable.

In Earman and Norton's argument, the static shift, and others like it, are best regarded as intuition pumps. As thought experiments they have far stronger appeal than Leibniz's risky premises. They have played a major role in our thinking about the reality of space – and spacetime. Daniel Dennett describes intuition pumps like this:

> If you look at the history of philosophy, you see that all the great and influential stuff has been technically full of holes but utterly memorable and vivid. They are what I call "intuition pumps" – lovely thought experiments. Like Plato's cave, and Descartes's evil demon, and Hobbes' vision of the state of nature and the social contract, and even Kant's idea of the categorical imperative. ... But they're wonderful imagination grabbers, jungle gyms for the imagination. They structure the way you think about a problem.
>
> (Dennett in Brockman (1995:180-97))

The shift arguments certainly grab our imaginations. We swarm over the jungle gym and swing toward the answer "No difference: it is superfluous to say that things have spatial relations to space as well as to other things." The shifts leave us in the same world. At first sight they are vivid, easy, compelling arguments. To change Dennett's metaphor, they dazzle us.

But must anyone answer yes to all or any of this? No one must; moreover, no one should. Arguments for this claim follow. None of them involves any reformulation of the substantivalist's position.

1.3 Shifts in spherical space: S_2 and S_3

Let us begin with the "even better" version of the static shift suggested by Earman and Norton: translate everything the same distance in the same direction. We will see that, in spaces that are non-Euclidean, the shift always misfires because these spaces have no parallels. It always backfires as well. The objection fails. There are many perfectly genuine spaces for which the shift cannot preserve relations among bodies. They are neither rare nor especially complex. None of the simplest non-Euclidean three-dimensional spaces yield a static shift symmetry that preserves relations among bodies. That is what I call misfiring. For example, in the space S_3, the three dimensional analogue of the ordinary two dimensional spherical surface, the shift does not even quite make sense, let alone yield what is claimed for it. Leibniz's aim was to show that space is a chimera. He could not have known that there were spaces other than Euclid's that resist his tactic; but that ignorance makes him vulnerable, despite the extraordinary ingenuity of his work. The quote from Earman and Norton comes in an argument against substantivalist views on the range of spacetimes in General Relativity. If the following objections are valid, the authors have overlooked the fact that the shift they describe does not work outside Euclidean space.

It is easy to see how by first looking at the shift in S_2, the familiar spherical two-dimensional surface considered as a complete self-contained space. This will allow us to take Earman and Norton's words in a literal sense, although not the sense they intended. They intend the shift East to be in the three-dimensional Euclidean space that contains the spherical earth. But East on earth points in no one direction in that containing space. To find any such direction, we must first specify it as a tangent to East-on-Earth, a tangent that points out into three-space. Then we need to say at which earthly point we take the tangent, and at what time, both of day and year, because the tangent direction in the containing space is different when these differ.

East and West are properly directions only on the earth's surface. But when we consider the global shifts in these directions in the self-contained space of the spherical surface, they don't work. The problems don't rest on the difference in dimension between the spaces S_2 and S_3. They depend only on very similar geometrical structures in S_2 and S_3 and marked differences from them in the structure of Euclidean space of two or more dimensions. So let us consider the simple case of directions on a complete two-dimensional spherical surface, pretending, for the moment, that the surface of the earth meets that ideal. Then it is easy to see what the problems are.

What would happen if all objects on the terrestrial surface were shifted the same distance East? The question is strictly nonsense. At the North Pole, every direction is south. Further, at, say 10 miles south of it, a shift of 20π miles due East will bring an object back to where it was[1]. That won't be the result if you perform the operation anywhere else – say 50 miles south. Shifting things a fixed distance East will change all the relations between bodies, the only exceptions being at places where the operation makes no sense.

East works differently from other terrestrial directions. Suppose

---

[1] The ratio of circumference to radius is a bit less than π even for a small circle on a sphere. But I guess the reader will see the point nevertheless.

instead, everything is moved 50 miles North. For anything within 50 miles of the North Pole, this is nonsense again. For places further away, symmetry fails in a different way: things that were north of the equator will get closer together in East-West separation, crowding acutely and overlapping near the pole: things more than 50 miles south of the equator spread out, the more so the nearer to the South Pole. Shifts specified as motions along latitudes are not like shifts defined by changes in longitude. Where the shift makes sense, it will change all the relations between bodies. The objection transforms easily and intuitively to S_3.

The acid test fails. Everyone should answer yes: the worlds are different. In general, the shift misfires. Worse, it backfires. That comes next.

Something more serious than my niggle about "East" goes wrong. Of course, there are indeed ways to get a shift-symmetry of the spherical surface (a symmetry is what is being fished for) but you can't get it with the shift described. For instance, you can keep things at North and South poles fixed and move all the rest together through the same longitudinal angle: that's a symmetry of the spherical surface. But things are not then moved the same distance East. Further, this shift, in turn, is meaningless on a Euclidean surface and in Euclidean three-space: they have no polar points. Only a precise instruction which operation is to be performed, and *in which space*, will determine whether it results in a symmetry, an asymmetry or a nonsense. Thus the shifts not only misfire; they backfire too. They show that thing-to-thing spatial relations *are dependent* on thing-to-space ones. More accurately, they show that these two classes of relations are not different and separable sorts of spatial relations: as *spatial* relations, they differ only in the *terms* that stand in them. There is no such dualism among spatial *relations*. We can talk sensibly only of definite shifts in definite spaces: only then is the question of the acid test answerable. Then the answer is clear but not what the shifter wants.

1.4 Other shifts

Now for nocturnal expansion. Turning to spherical space and its visible analogue, the surface S_2, again quickly gets even beginners to see that space can have a geometry in which the effects of change in size of things in the surface is plainly discernible. Just imagine trying to double the areas of earth's oceans in the space of the earth as it is. Equally clearly, in spherical 2-space, changing size changes shapes. Differently sized triangles drawn on the surface quickly illustrate that. *The structure of the whole ambient space determines which spatial relations can obtain among things.* The supposed duality among spatial relations fails again.

This puts us in a position to see that Leibniz's "quite the contrary way by... changing East into West" fares somewhat better than these later kinds of shift, despite its ambiguity between reflection and rotation. Yet it fails in a different and perhaps more illuminating way. The real space of the surface of the earth is not strictly symmetrical. It is an oblate spheroid. You can't rotate a flattish mid-latitude city like Adelaide onto either of the polar regions (they're too flat) or equatorial ones (too curved) and retain the relations among its streets and buildings. That fails, too, where the surface is locally irregular - for instance, in the Andes where local curvatures are both sharp and varied. You can't rebuild Machu Picchu anywhere else. Never mind architects' and builders' problems. The geometry of the terrain will be wrong. Reflection yields symmetries only if the space has them.

Consider, now, something rather different: a Holus Bolus shift. When you double the size of things, double the space as well. That's always a symmetry. Actually it need not be[2], but the main point is that Holus Bolus shifts are irrelevant. To claim that space is superfluous, the shifts must show that we can detach thing-to-thing spatial relations from thing-to-space ones and freely re-attach them anywhere, leaving the real state of the world unchanged. But in Holus-Bolus shifts, things remain attached to the same points. That is how they guarantee symmetry. So it has no power to show that thing-to-thing relations are detachable, so no power to defend the spurious dualism among spatial relations.

To sum up: these shift intuition pumps empower two false theses (i) the detachment thesis, that thing-thing spatial relations may always be detached from thing-space relations, then re-attached unchanged (ii) the Principle of the Identity of Indiscernibles. The Principle is irrelevant at this stage, insufficient at any stage, and implausible in itself. It will be more closely scrutinised in §4.

This unmasks the core strategy of shifts. They aim to show that spatial relations of things to things can always be *detached* from spatial relations of things to space and freely, arbitrarily *re-attached* without change in the former relations among things. Grant that symmetry and the Principle of the Identity of Indiscernibles – that no two things or states of affairs have all their properties in common – is then (contestably) invoked to conclude that the states of affairs before and after the shift are identical because indiscernible. Then thing-to-space spatial relations are detachable from thing-to-thing relations and may be thrown away as superfluous metaphysical junk. Space goes out with the garbage, too. That presupposes that the only thing-to-space relation that matters in this context is *being at*.

---

[2]The point is discussed in detail in Nerlich (1991).

1.5 Detachment and a dualism among relations?

This *detachment thesis* insinuates a dualism – that differences in the ontic type of *terms* that stand in spatial relations yield two different, independent kinds of spatial *relatedness*. There are, first, spatial relations of things-to-things; second of things-to-space. The first relations are widely regarded as carrying no commitment to space or parts of space[3]. Somehow the relation is "direct" while remaining spatial. This thought is the child of the Leibniz shifts. But it is not clear what it means to describe these relations as spatial once detached from space itself. The second group of relations, those that separate and connect parts of space, are regarded as committing us to spatial paths to do the separating and connecting. This hint of dualism has shaped the long and dominant debate between relationists and substantivalists. Relationism accepts the Leibnizian conclusion: crudely, there are only spatial relations of things to things: the rest is metaphysical rubbish. Substantivalism, as usually described, has defended the reality of space (spacetime) but has not challenged the claim that the thing-to-thing relations' carry no commitment to space or its parts. Instead, substantivalists stress a *further* need for parts of space and the second sort of spatial relations among them: only that sustains adequate spatial discourse. Its main tactic lies not in disputing the relationist thesis that there are "direct" thing-to-thing spatial relations but in mounting counter-examples to the claim that they are adequate by themselves. Thus both theories accept the shift arguments and ignore the question what the nature of spatial relations among things actually is. So both work within the context of the dualism just described. But if the earlier objections to the shift arguments are sound then *there are no spatial relations independent of what space they are in*. That is why the shifts always backfire. Thus it is important to look for a theory that sees the same relativity in all spatial relations. Realism is the broader view that there is no sound revision of the form of spatial discourse that purges it of commitment to space. Substantivalism is just one, and a dubious, form of realism. This will be discussed in detail in the next chapter.

1.6 On the importance of regions

Leibniz's opening sentence is misleading in its second part – "one point of space does not absolutely differ in any respect from another point in space." However, nothing about the sameness or difference in the

---

[3]If they did the debate would be settled at once in favour of realism.

geometry of regions follows from properties intrinsic to their points. The shifts fail because the structure of regions or of whole spaces can differ significantly from one another: in curvature for instance. Someone might object that curvature is an intrinsic property of a point; if so, points may differ in their intrinsic properties. But curvature really belongs to a point as the limit of a structural property of regions that contain it. The failure of the shifts does not depend on properties of points but rests on the structure of regions or of the whole of space. Thing-to-thing relations within a region are merely parts of a structure of relations that link all spatial parts of a region to each other.

Of course, none of the quoted authors really meant to talk about surfaces, not even a spherical one. They aimed at three (or higher) dimensional spaces[4]. That cannot save the acid test. The *particular* moral of the counter-example that takes "East" and "West" literally, is that the peculiar geometry of complete spherical spaces of any dimension, differs widely in the structure of its regions from that of Euclidean spaces (of any dimensions). The general moral is that shifts are intelligible only when a matching space is specified. It makes no sense to think of thing-to-thing spatial relations other than as parts of the whole network of relations that make up the structure of a region of space. The acid test, as it was intended, presupposes that Euclidean geometry is the only geometry. No one, not even a substantivalist, should agree with that.

1.7 The unnatural longevity of the shift arguments

The longevity of Leibniz Equivalence and especially of the shifts is surprising and puzzling. In the case of Earman and Norton the mistake does not really lie in their use of "East" as if it were a direction in three-space. It's their apparent failure to see that the intended "better" shift presupposes that, at any two points in any space, the same direction of shift is well defined. This fails in all examples of the quite simple and straightforward classical non-Euclidean spaces. Thus shifters must presuppose the "right" space or the shift instruction fails to make sense or fails to entail the right outcome. It is surprising that Earman and Norton, who know perfectly well that space need not have the parallelism of Euclid's space, should argue in a way that presupposes it. They are writers in the front rank. Conceivably, Leibniz Equivalence is so entrenched in our tradition that *rethinking* the argument may not cross even the most ingenious minds. However

---

[4]No doubt they mean a direction in Euclidean three-dimensional space. But, as argued above, no one such direction is simply east.

suggestions that might explain the phenomenon seem rather implausible and rather impertinent. The Hole problem itself, while it exploits Leibniz equivalence, is a more complex matter;[5] it has been probed by many distinguished philosophers of science and it goes on inviting attention. No contributor to that debate hints at any problem with the *shifts* or the analogy with Leibniz.

Further, the misfiring of the shifts has long been well known. The very first thought that led to the non-Euclidean geometries was that the parallels axiom is not self-evident. It makes a claim about infinite space – that parallel lines never meet. How could we know that? An immediate result of dropping it is that the same direction at different points is not uniquely defined.[6] That obviously equates to the absence of parallels. It was soon found, also, that the only similarity geometry is Euclid's. That is, among classical non-Euclidean geometries,[7] none admits figures with the same shape but different sizes.

In a prominent paper, Helmholtz (1960)[8] coined the phrase "free mobility," for the rigid, undistorted motion of bodies that is possible only in the special group of spaces with constant curvature. He, along with other of his contemporaries, argued from this that such spaces, homogeneous and isotropic, are the only candidates for the actual space of physics since physics was thought to depend on the possibility of rigid bodies. This famous work established that free mobility – i.e. translation symmetry – is a highly non-trivial feature only of those spaces. So it is well known that there are spaces where the Leibniz Equivalence fails.

1.8 Who may exploit the shifts?

It is widely assumed that the shifts are weapons in the relationist's arsenal but, on the contrary, they are guns for its opponents. Here's how. Shifts, in general, will not yield symmetries. Some will, but only within appropriate, sufficiently symmetric spaces. Symmetry is a necessary condition for an appeal to indiscernibility and thus to the identity of states of affairs before and after the shift. When symmetry fails it is because these states differ. That brings into play the overwhelmingly more persuasive converse of Leibniz's Principle of the Identity of Indiscernibles, the Principle of the Diversity of Discernibles (Indiscernibility of Identicals):

---

[5]See Chapter 10.

[6]It can be defined only relative to a path between the points.

[7]I mean the geometries of constant curvature, sub-geometries of projective geometry.

[8]Helmholtz was first but by no means last to note free mobility as an important restriction on the range of geometries that permit physical measurement. It played a part in Russell's geometry, among others. See Torretti (1978:3.1.1-3; 4.3.5).

$$\text{If } Fa \text{ and } \sim Fb \text{ then } a \neq b.$$

But if the relational states before and after a shift are really different, as they well may be, the property or relation that makes the difference is real, too. In our examples, it's the relations of things to their containing spaces that explain how some shifts yield plainly different worlds. Something real, what the shift is a shift *within* or *through* or *into*, makes a real difference. It's space. So space is a crucial factor in any thing-to-thing spatial relation, not a superfluous chimera. Shifts without symmetries don't just misfire on the relationist: they backfire!

Shifts are not the quick of this ulcer; spaces are. Some spaces admit a symmetry with the right kind of shift (transformation). But we've been using the wrong shifts in the wrong spaces. A shift that will work in Euclidean space won't work in spherical space. A shift that will work on the sphere doesn't even make sense applied to the plane. Shifts mostly misfire. Without a symmetry, there is nothing on which the Principle of the Identity of Indiscernibles, the star actor waiting in the wings, could work its wonders. Further, shifters need an argument that will apply to the concept of space *quite generally*. It's no use hoping to show merely that this or that geometry is a chimera – and we are still a long way from seeing even that.

We saw that, in some spaces, those with variable curvature, no shift is a symmetry. On the other hand, in Euclidean space, clearly, many shifts retain thing-thing relations: it is no matter whether it is a rotation, translation or reflection. That draws attention to misleading emphases on describing the shift in common versions of the argument: first, what is required is that thing-to-thing replacement with symmetry is possible, never mind by which shift; second, the preoccupation with specific shifts may lead to omitting a crucial requirement: relation-preserving replacement must be possible *anywhere at all* in the target space. There can be no question of passing over unsuitably curved regions to settle things somewhere more welcoming. The whole space, everywhere, must allow the symmetry. That is, the space must be isometric throughout. We might now speak of a Leibniz "dump" to play down the prominence of shifts – replacement anywhere, never mind how. Specifying shifts invigorates imagination but distracts from vital issues. However one detaches the thing-thing spatial relations he must be free to re-attach them anywhere – just to dump them.

1.9 Homogeneous spaces and shift symmetries

The Principle of the Identity of Indiscernibles has yet to play a part in this critique of the shift arguments since no symmetry has cued its entry. Yet some shifts in some spaces certainly result in symmetries. In homogeneous and isotropic spaces quite generally thing-to-thing relation symmetry is always available. What happens when a shift meshes with the gears of the Principle? Without loss of generality we may consider just Euclidean space. Let's try again with "translate everything the same distance in the same direction." This particular shift clearly meets two necessary conditions. It makes sense in Euclidean space and it yields a *symmetry* there since the same direction can be everywhere uniquely defined. The acid test question asks whether the original and shifted worlds are different? To say they are is to reject the Principle.

Substantivalists have no special commitment to the concept of Euclidean space although it would be surprising if any were to reject it. But our critique of the shifts should caution anyone to ask exactly which question the shifter is asking. As a presupposition of the shift's intelligibility, does the *shifter* insinuate that we are to consider precisely Euclidean space? Realists, aware that shifts don't automatically yield symmetries, should proceed cautiously – just what does the shifter presuppose? Shifters neither limit the spatial spread of objects in the universe to a finite region, nor constrain how big a shift they want us to consider. They just pluck some distance out of the air. Thus realists should insist on its being explicit whether or not they are invited to consider an *infinite Euclidean* space: do shifters pose the question in the context of an infinite space with infinitely many indistinguishable parts?

If so, the outcome of the explicated shift is clear: different indistinguishable world-states result. Substantivalists should answer yes to the acid test question in this case and so should everyone else. The *shifter* sets aside the Principle of the Identity of Indiscernibles in asking the question, not substantivalists in answering it. The shifter is in no position to charge them with the folly of countenancing indistinguishable states of affairs. That was countenanced when the shift presuppositions were made explicit.

If the Principle of the Identity of Indiscernibles were true, then the first response should be that the *shift instruction* is a senseless chimera. But should Euclidean space really be deemed devoid of sense on the ground that it has infinitely many indistinguishable parts? The shifts have been taken to illustrate the Principle's power as an ontological tool. They backfire. Even without the varieties of geometry, the Principle is irrelevant to the shifts without a question begging, and now plainly false, assumption that thing-thing spatial relations earn a privilege over thing-space ones. Without relations to an ambient space there is no definite question posed by a shift and no definite answer.

The Principle can be phrased so as to be explicit, certain and trivial. If we include "...is identical with..." among the properties that are allowed to have a discerning role then, indeed, if A has all the properties that B has, and that includes being identical with B, then "A is identical with B" will be a trivially valid consequence. But that understanding of the Principle is of no use in the meatier cases that worry us. Often it is formulated foggily enough as to lack these desirable properties. It may tacitly insinuate, as if it were the meaning of the Principle, a negative judgement on some property that it justifies dismissing as making any difference. Indeed we saw just this happen in the shift argument. Then the Principle is obscurantist, posing as a justification of what it tacitly assumes. For the shifts to work, we need to gloss the Principle, so that the property "...is at a different place from..." is disallowed as a discerning difference. But once the Principle presupposes this highly non-trivial rule in its application, it begs our question.[9]

In fact the Principle is seldom used so that it is explicit which properties are supposed to discern (or fail to discern). Alternatively it may be stated in so specific a form as to have a very restricted application. (Norton, 2011:§5), in his limpid and impressive article on the Hole Argument, states Leibniz Equivalence as follows:

> If two distributions of fields are related by a smooth transformation, then they represent the same physical systems

The welcome feature of this statement is that it tells us which property is to be discounted as making a difference between the distributions. It is silent on any general statement of criteria for identity or difference. So it is far from the Principle as Leibniz would have expressed it. It does not look like a doctrine in metaphysics and it is too specific to cover all the examples that Norton needs – the shift examples in particular. They are not about fields and smooth transformations. Overall, in Norton's development of the Hole Argument, the Principle is rather a chameleon. Further, Norton (op. cit. §8, second last paragraph) at least *appears* simply to waive it when its consequences are unwelcome, and that in a case close to the shift arguments. In the context of the Hole, it looks nothing like a Principle of metaphysics, and far more like a dressing up of eminently sensi-

---

[9]In current usage, the phrase "begs the question" increasingly has the sense "...raises the question" rather than the sense long dominant in philosophical prose "...begs off answering the question." I stick to that honourable, I hope not vanishing, usage.

ble but purely pragmatic decisions that are of almost no philosophical interest. An account of that must wait till Chapter 10.

In variably curved spaces the Principle of the Identity of Indiscernibles has no symmetries to work on: none of them can be unmasked as a chimera. Do the symmetries of Euclidean, and other homogenous spaces, really entail, via the Principle, that they *alone* among possible spaces are senseless monstrosities? Endorse Leibniz Equivalence and that conclusion follows! Euclidean space has been a darling object of claims to necessary knowledge and perfect understanding for over twenty three centuries of geometric thought. It is far more plausible that Euclidean space is intelligible than that the Principle of the Identity of Indiscernibles is true (to say nothing of the implausible Principle of Sufficient Reason). If anything here is an inchoate chimera it is Leibniz Equivalence.

Perhaps the reason why Leibniz overlooked backfiring is that he argues (as quoted above) that the Euclidean symmetries arise from the indiscernibility of one *spatial* point from another (analogously for temporal points). But differences in spatial geometries are perfectly compatible with the indiscernibility of points. It is regional differences in spatial structure – differences in curvature – that give us the infinite range of cases in which shifts misfire. From the perspective of points, differences merely of place among the relata (the points they are *at*) can't affect the structures of the relations among them. And, indeed, what counts is the difference among *all* the relations of *all* the relata in some spatial region. Spatial relations among things are simply part of the spatial structure of the whole region, local or global, that they are in. What matters is not the places that things are *at* but the structure of the wider spatial regions they are *in*. That is why the shifts misfire and backfire and that is why we can't detach thing-thing relations from thing-space relations and throw the latter away.

I have not advanced substantivalist arguments here. Substantivalists are rare birds at best, since substantivalism is largely described by those who, like me, wish to distance themselves from it. Substantivalists mistakenly accept that spatial relations among things do not involve parts of space - connecting paths and the like. They accept that the shifts lead to thing-to-thing symmetries that are independent of thing-to-space relations in that respect. They accept the dualism among spatial relations. Substantivalists fail to argue that spatial relations among things can't be considered apart from their membership in a wider class of spatial relations. It is a form of spatial realism: however, realism is not a form of substantivalism.

Finally, if GR is true we live in a variably curved spacetime. A "natural" choice of space in our environment as in Schwarzschild coordinates for our roughly Schwarzschild local spacetime yields a variable curvature of both space and time. The spatial curvature may not be observable but the theory tells us that it is certainly there. Further, what's observable changes with advancing technology.

I conclude that the acid test, as tied to Leibniz's shifts and as described by Earman and Norton (op. cit.), is harmless to substantivalists or to any other kind of space and spacetime realists. A main thesis in this conclusion is that spatial relations among things are never independent of the geometry of the region which the related things are found in.

2 Path Realism: The Traditional Debate Reshaped

Chapter 2 examines the traditional debate over the reality of space between relationists and substantivalists. It proposes a way between the horns of that dilemma. Spatial relations among observables are fundamental but only epistemologically. Both traditional positions explicitly or implicitly admit the shift arguments. The way between begins with spatial relations among just observable things. These are external relations. It is argued that separating and connecting paths make them external. Both spatial and spacetime paths are argued to pervade outer sense perception. Since paths are spatial or spatiotemporal things, their fusion is space (spacetime) itself so relationism cannot admit them: it fails to have an account of externality of these relations. It is argued that the standard definition of substantivalism does not apply to path realism. Paths directly relate things; they are concrete without being material; so, therefore, is space. Relationism's problem with reflexive spatial relations is explored.

2.1 The traditional debate

The failure of Leibniz equivalence requires a reshaping of the traditional debate between spatial relationists and substantivalists. A rather different player in the game can best exploit the way the shift arguments backfire. We need *path realism* or, for short, just realism. "Substantivalism" also names a form of spatial realism but is regularly used, misleadingly, to cover all views that take space as a real thing. Like relationism, path realism begins as a theory of spatial relations among objects; in the end, it is intended as a model for temporal and spatiotemporal realism, too. Spatial and temporal relations among observables certainly play a fundamental role in the epistemology of our understanding of space and time. Spatial (and temporal) relationism indexrelationism!temporal defends the adequacy of just those spatial (temporal) relations that hold among observable things (events): these relations equip us with all we need for discourse about space and time. The rest is objectionable on both epistemological and metaphysical grounds. Substantivalism's core thesis is just that thing-to-thing relations are not adequate. It neither challenges the shift arguments nor argues that thing-to-thing spatial relations already depend on the spatial structure throughout regions or of space (time) as a whole. By contrast, path realism rests on that thesis. It has only a secondary concern with the wrangle over adequacy. Its main arguments are simplest, clearest and deepest when they begin with thing-to-thing spatial relations.

However, relationism and substantivalism have shaped the traditional debate about the ontology of space by the way they respond to shift arguments. Here are two representative definitions:

> *Substantivalism*: "Space-time is a substance in that it forms a substrate that underlies physical events and processes, and spatiotemporal relations among such events and processes are parasitic on the spatiotemporal relations inherent in the substratum of space-time points and regions."
> (Earman, 1989:11)
>
> *Relationism*: "Spatiotemporal relations among bodies and events are direct; that is, they are not parasitic on relations among a substratum of space points that underlie bodies or space-time points that underlie events."
> (Earman, 1989:12)

There are other definitions of these terms (Lucas (1984:191-195) gives many). Earman's relate directly to what is presently my target and they are rightly regarded as authoritative statements of the main views in the traditional debate. The view that space (spacetime) exists independently of its contents is also laid at substantivalism's door (Norton, 2011).

2.2 What makes spatial relations spatial?

The main issue has been whether thing-to-thing spatial relations alone encompass all real spatial facts and features. That ignores a prior, deeper question about all spatial relations – what makes them *spatial*? In the state of affairs that makes a statement about spatial relations true (its truthmaker) what is the condition that makes it spatial rather than, say, temporal. It must include something more than their terms – more than the properties intrinsic to the things it relates. That goes for temporal and spacetime relations, too.

Perhaps that is obvious, but objections arising in discussion suggest a need to add more. Here's a much-cited paradigm of relationism in action written by the great mathematician, Hermann Weyl. The issue in contention is Kant's claim[1] that, if a hand were to exist in a world where it was the only object, it would either be left or right, not indeterminate in handedness. Weyl objects as follows

> Had God, rather than making first a left hand and then a right hand, started with a right hand and then formed another right hand, he would have changed the plan of the universe not in the first but in the second act, by bringing forth a hand which was equally rather than oppositely oriented to the first created specimen.
> Weyl (1952:21)

Much is cryptic in these remarks. To clarify: "The second act" refers to the creation of a second *hand*, not to the two *plans* of creating congruent or incongruent hands.

Weyl's complaint is that Kant gratuitously assumes a space surrounding the hand. For relationists, that assumption in this context is unintelligible. There is no difference between beginning with a left hand and beginning with a right: only choices in the second act make the plans different. That is a deduction from relationism, not Weyl's argument for it. Relationism entails that all the spatial relations there are hold between parts of the lone hand; they are all internal to the volume exactly filled by it. Thus there can be no outside space to move the hand through or to locate its reflection in. It is neither left nor right nor handed. This is a *counter* challenge to Kant's.

However, Weyl himself makes an assumption: the existence of a second hand immediately provides a structure necessary and sufficient for handedness. The assumption is false.

If God's second act of creation is exactly the same as his first, it, too, creates a hand without external spatial relations. Weyl assumes that the second hand, somehow, just in itself, yields spatial relations. He needs it to yield something more than just another hand else there could be no difference between creating incongruous rather than congruous counterparts. Weyl says nothing about what this something else is. So it is gratuitous, just as Kant's assumption was.

Formally the outcome of the two acts, if the second strictly repeats the first, is this:

$$(\exists x) (x \text{ is a hand}) \ \& \ (\exists y)(y \text{ is a hand} \ \& \ (x \neq y))$$

---

[1] See Van Cleve and Frederick (1991:27-28).

Nothing entails that either hand instantiates any external relation to anything outside itself. Weyl assumes a structurally richer world than he explicitly describes. Since relationism regards creating a second hand as ontologically prior to an external spatial relationship, the relevant creative acts must be structurally richer than the creation of the hand. Weyl's god must create external spatial relations as well as a hand. He must create paths.

There are three different possible worlds. W_1 has two hands spatially unrelated; in W_2, the hands are externally related as congruous counterparts; in W_3, as incongruous ones. The worlds differ in structure. What structure? It is not that relationism provides the wrong answer to this question: It provides no answer.

Path realism admits only direct spatial relations in the truthmakers of all spatial statements whatever their terms. It needs no indirect relations or parasites on hosts. The path that links x to y is the same as the one that links the place of x to the place of y.

Lawrence Sklar writes:

> ... attempts to define the "spatiality" of spatial relations are usually taken as ineffective or unintelligible, so I will simply give the relationist his designated family of relations.
>
> Sklar (1974b:169).

Sklar frankly bypasses the prior question, endorsing only one side of the (spurious) dualism among spatial relations. But the reason why the definitions are ineffective or unintelligible is just that they have been detached from the space that made them spatial in the beginning. There is no way to define their spatiality short of linking them by paths thus putting them in some space. But unless we know which space, we can't know whether the newly attached relations among bodies can ever be the same, under a shift, as the old ones. The space has to be the same, too. The shift failures show that globally shifted things cannot automatically carry their spatial relations with them. Things will not wind up similarly related except in special cases of global symmetry. The thing-space relation "x is at y" is too weak to reveal how spatial relations among things depend on the relations of the thing to space. That is a regional or global matter.

In fact, Sklar gives a decisive reason for *refusing* to hand over the family of relations. He simply assumes that there are only space-independent thing-to-thing spatial relations at issue. If that were so, it would indeed follow that they must be the property of relationists rather than substantivalists. Without that assumption, the reason for handing the relationists "their" relations lapses and becomes a conclusive reason for withholding them. Sklar is explicit: relationists have not been able to say what spatial relations are. He doesn't presuppose a relationist answer – he confesses that there isn't one. The ineffectual attempts he mentions are Leibniz-like efforts to *reduce* spatial relations to non-spatial properties of the things they relate, or derive them from other external relations, cause being most common. That's why they are unintelligible or ineffective. Sklar's judgement on relationist's relations is correct: relationists have no good theory of them and can't have one until they are somehow explained as spatial. Till then, it is not clear just what it is that Sklar gives them. In particular, it is not clear that the relations are not, after all, path-realist spatial relations. To describe them is our next topic. Once relationism accepts Sklar's unspecified gift, it risks becoming realism-in-denial.

2.3 An entrenched belief

The core issue is simple. It is an entrenched belief about spatial relations that nothing intrinsic to an object tells us where it is, whether there are other objects anywhere, how distant it is from any other, whether it is between two other objects and so on. Conversely, that two things are spatially related puts no constraint on what they are, unless they are necessarily non-spatial (numbers for instance). Spatial relations are not grounded in the natures of their terms as internal relations, such as similarity, are. They are *external* relations. This is clear in the particular case of distance. It is a symmetric relation, but, unlike internal symmetric relations, it has nothing to do with the intrinsic properties of the things it relates. So what grounds this relation, and what grounds other, non-metrical relations – e.g. ternary ones like "between"? How are they *spatial* relations?

The entrenched belief plausibly allows just two options. First, spatial relations might be definable in terms of non-spatiotemporal properties. Yet it is not clear what we have to work with unless the definition involves intrinsic properties. If it does involve them, the relations will not be external. Second, while external causal relations have been front-runners in this race, they have never breasted the tape. All bog down in one or another quagmire of the sort Sklar gloomily indicates. That is partly because cause is not a purely external relation. It depends on which qualities in the cause produce which qualities in the effect. Further, cause has always been a strongly contested notion. Last, cause presupposes spatio-temporal relations and does not ground them. There is a third option: spatial relations are mediated relations, i.e. their truthmakers include an *entity* linking the relata: it is not a property, and so is independent of the terms and their natures.

The negative message of relationism is clear: there are no unoccupied points; spatial relations among spatially separated things are not grounded or mediated by paths in space. There is no positive statement about what grounds spatial relations.

Relationists usually take it that they need no account. They base this on the Leibniz shift arguments criticised in Chapter 1.

Some relationists propose possible objects to mediate spatial relations among separated things (Sklar, 1974a). No actual entity is needed. An analysis of a paradigm sentence for unoccupied distance runs: *Possibly [3 metre-rules laid end to end span the interval from a to b]*. The difficulty here is to identify the sense of the operator "Possibly." It is *logically* possible that 3 metre-rules span the distance. But that is so no matter how far apart the things actually are, and true even if they are not in the same space and are not at all spatially related. It is also possible relative merely to physical laws, no matter how far apart the spatially related things are. Question-begging identifications, especially that the sentence is possible *relative to the fact that the two things are 3 metres apart*, must be avoided. It is unclear how to do this. (This objection is further developed in Nerlich in Van Cleve and Frederick (1991:166-70); Earman (1989:167-8).)

Mediation, restricted to occupied paths, does not help. It is usually presupposed that the terms of the relation will be material points in a matter or field continuum. Simple materialised extension is not enough. Continuity is required to permit the use of differential equations. It is doubtful whether material points (parts) are suitably related to each other spatially to meet the continuity (or any other) condition. Field distributions are continuous but parasitic on spatial continuity. Those relations are just the ones in question: we are back at the problem we began with. Touching may be offered as intrinsic to things that touch. It is doubtful that this succeeds (see Hooker (1971)).

2.4 On primitives

Since intrinsic properties of terms don't do the work, the obvious mediators are described, pretty well synonymously, as paths, separators-and-connectors or spatial relaters. Space is their fusion. "Path," like "space" and "time" themselves, is a primitive idea – too familiar and basic to admit definition in other terms, yet clear, familiar and simple enough to be described and identified as spatial. Compare Newton: "I do not define time, space, place and motion [since they are] familiar to all" (Newton, 1999:Scholium). By contrast, it is no good making a primitive of something merely because you don't know what it is – as Sklar does, in effect. It is a worrying thought that the question of adequacy – are spatial relations among observable things sufficient to sustain spatial discourse? – is often considered without any good description of what the relations actually are. Primitiveness may be vulnerable to new accounts and definitions based on them. Just as there may be scholia to definitions, telling us about the thing mentioned without defining it, so we need scholia to let us know something about what our primitives are. Newton limned the essences of space and time, not in the Definitions, but in its famous sequel, the Scholium. His essentialist style may repel, but his explication of primitives is exemplary.

Paths are primitive but we know what they are since they pervade our perceptions. We can show what they are by pointing them out.

2.5 Paths and perception

"Path" is a familiar word and paths are familiar in concrete perceptual experience, separating and connecting things. We can immediately distinguish two objects, exactly alike to perception: they are separated yet, since seen together, connected. True, we do not see separating, connecting paths; they aren't visual objects. Yet we do immediately see *that there is* a path. Indeed, we will invariably see that there are many. We need not see any occupant of them. We don't see or feel paths but we see through them, move and look along them, look across them, trace them and so on. They are among the commonest entities encountered in concrete, immediate perception and action. Although we are confining attention just to relations among objects, our perceptual path-tracing skills do not depend on tracing objects. For example, there is no pole star for the southern hemisphere, but there is a rough night-sky recipe for southing. Follow the long axis of the Southern Cross southwards beyond the constellation itself. At the south rotational pole of the sky, it intersects the perpendicular from the mid-point of the line joining the "pointer" stars. It's an easy straight-path-tracing recipe. Beyond the four mentioned, no intervening stars are seen. No object is seen to move along the paths. These paths are concrete entities, familiar in perception.

In fact we see spacetime paths mostly without being conscious of it. Toss a ball from hand to hand, watch it through its rise and fall and you perceive its spacetime path. It takes a little time to perceive a thing move along a timelike path. But we regularly anticipate such paths before a moving object has marked them out. Of course we learn this skill by watching things traverse paths. But cricketers, tennis players and so on know quite well how to throw or strike a ball so that it will travel to a planned place arriving at an anticipated time. That happens when a bowler beats a batsman for pace as commentators say. Players know where and when to move in order to take a catch, return a stroke and so on. It's not just spatial but spacetime paths that we can, familiarly and well, judge before anything fills them.

Observers in different frames of reference see the same path but in different ways.

We are poor judges of the distance from ourselves of two stars seen as separated and connected, and we misjudge the linking path's orientation in spacetime. Perceptual immediacy does not entail infallibility. We naively think we see the stars as linked by a spatial rather than a spacetime path, since we don't see that the light from one is from an event much earlier (in our frame of reference) than that from the other. We see stars as at spatial distances but what we see is also at a distance in time. Further, the two are at different distances in time even though we never see any events *as past*. At any star-gazing moment, we see events of light emission from stars – call them star-events for short. However good our judgements are in everyday terrestrial gazing we are poor judges of what we see in everynight star-gazing. Looking at a star-event along its light-path we go wildly wrong in how far away it is in space and time. But, in the southing recipe, we are in search of a *direction* rather than guessing the size of a spacetime interval. What we trace out, following any one step in the recipe, is a light-like angle from the star-event in spacetime of one star to the star-event in spacetime of the next. Both are on our past light cone. The angle swept out subtends infinitely many paths in spacetime only one of which links the spacetime point of one star-event to the other. Our gaze sweeps across, yet we can't pick out, the single spacetime path linking the star-events. By sweeping out the angle we trace all the straight paths in the spacetime surface it subtends including the one that links the emission events. If the stars themselves endure long enough, they are separated by both spacelike and timelike paths that connect points in their histories. We see through and across all these paths; paths do not mask other paths, although, in these circumstances, we lack the power to distinguish them. They are perceptually immediate; there in our visual fields. However, they are epistemologically remote. We don't know much about them.

At the end of this tracing process we get what we wanted: a new direction, the South rotational pole of the sky. That is to say we locate and look along a null (or lightlike) trajectory through spacetime, each point of which is due south in the terrestrially based coordinate system with which we map the heavens.

Relationists may claim the empiricist advantage that their view is close to observation since the related objects are observable. But we also observe that states of affairs exist. While the nature of the spatial relation remains unspecified, it is not clear just what is observable in the states of affairs e.g. that x is near y but not near z, that z is between x and y and so on. Realists insist that their view is close to observation since paths are visually traceable. It is useless for relationists to point to examples of the relation, since whether paths are part of what is pointed at, in indicating a state of affairs, is just what is in dispute. The need to include the path as well as the objects in the observed state of affairs is clearest, perhaps, in the thought that one easily sees the separation and connection of two things that don't differ perceptibly in their intrinsic qualities, if they differ at all. A structure is perceived beyond the things themselves. The relationist should describe it.

Relationist relations are often treated as uninterpreted primitives in a formal system. For instance, they obey the axioms for metric space, as realists agree they should. But the axioms fail to characterize the relations as spatial, even if they are enriched.[2] An interpretation of any such formal system is lacking. To supply it is to say how the relations are mediated.

Leibniz said of symmetric spatial relations that they are properties "with a leg in each substance" but a standing in neither. This pungently expresses the difficulty. Van Cleve (1991:213-7) has more on grounding. See Pooley (2003), Hoefer (2000), Huggett (2000), Huggett and Hoefer (2006) for recent relationist arguments.

If that is correct then what Sklar gives the relationist is, indeed, a family of *realist* relations. Paths make the relations spatial (spatiotemporal) since all spatial relations commit one to paths i.e. to parts of space. So relationism's claim to the *adequacy* of thing-thing spatial relations is beside the point of whether there are real spatial entities, since all of them need paths to be spatial relations at all. The classical debate, misled by the dualism of Chapter 1.5 ignores the core issue. All a realist need die for is that things are linked by paths. To see this, just look! We also see that many paths link two things only some of which are straight. Realism need not yet endorse any particular metric or topology, let alone the continuity and differentiability of space; nor need it tie itself to the doctrine that space could exist without objects. Although substantivalism is almost always stated as a doctrine about points, neither metaphysics nor simple observation commits us to them. Realists need not claim at the outset that paths are strictly 1-dimensional. These are theoretic *alternatives* for realism. Some of them look plausible.

---

[2] For both the constraints and their inadequacy, see Patterson (1969:28).

2.6 Distance without paths

A serious objection to this Gaussian[3] perspective on paths comes from the idea of a metric space. A metric space, in mathematics, is a formal object, a set of ordered pairs, the first member being a pair of "points," the second a non-negative real number. A standard interpretation of metric spaces casts the real number as the measure of a distance between pairs of spatial points, so it is required to satisfy the axioms for distance. It may also be used in other interpretations for which distance axioms are formally appropriate.

It has been argued (Bricker (1993), Dainton (2010:Chapter 9.6) among others) that this conception allows us to dispense with paths. The distance relation is unmediated. Then it justifies the distinction in the earlier definition of relationism that spatial relations hold directly between objects if we interpret "point" as "object." Since the relation in the standard interpretation is not specified as holding among objects, it is not strictly a relationist interpretation as defined earlier (Dainton (2010:160); see Grünbaum (1973) chapter 16.2 for a detailed study of intrinsic metric).

This contrasts with a Gaussian view of distance as the length along a path from P to Q, a shortest (or extremal) path giving *the* distance between the points. The Gaussian view accommodates differentiation and integration and elegantly yields concepts crucial to GR such as geodesic, vector transport and curvature. The theory works with the Gaussian concept of distance, not the weaker metric space concept. Since GR admits the idea that two points may be connected by different geodesics of different lengths, the approach via metric spaces would be, at best, awkward.

In order to know what we are talking about in different interpretations of metric space we must ask what the real number *numbers*: the answer may be something like this: It numbers the electric (gravitational etc.) potential difference between the points. For this it needs units and the usual background for them in terms of standards (of constant length or whatever). In short, it cannot be a pure number. In the spatial case, of course, it numbers the distance between them. But that is consistent with its numbering the length of the shortest (or an extremal) *path* from one point to the other. So to make this clearly not a path-realist move in the game, a non-path-interpretation must be given.

Bricker (op. cit.) tackles this in a particularly forceful and ingenious way by means of a thought experiment. I take it as appealing to

---

[3]In 1827 Gauss considered the problem of the metric of curved spaces, introduced coordinates for them and paved the way for the work of Riemann. Distance, as length of paths, was a core element in these new branches of geometry.

intuition or gut feelings, as Dainton (loc. cit.) puts it. But perhaps a lot comes down to these feelings at some stage or another.

First, imagine P and Q to be points in a real space with a path (extremal) between them. The number assigned to the pair $\{P,Q\}$ measures the length of this path. Since space is something real, it seems to make sense to suppose a hole could be cut in it so that a part of space that contains all or some of the shortest paths from P to Q, is removed.[4] Now the shortest path in the space must skirt the hole and be longer than the shortest path before. Yet, intuitively, P and Q are no further distant after this surgery. I also find that intuitive since it seems to bear both a clear relation to the distance along paths from P to Q that skirt the hole; also, if P and Q were moved in parallel along lines orthogonal to the shortest path that used to join them, they would be at the same shortest path distance as they were.[5] Bricker's thought is that there are two conceptions of distance here, one Gaussian and realist in my sense, the other path-independent as in metric spaces.

One might perform more drastic surgery: remove nearly all the space, leaving P and Q, each in its own distinct surrounding patch of real space. Now there is no path at all between P and Q. Yet it may still seem intuitive (not to me, I must say) that the surgery has left a distance between them, the same distance as before. That intuition rests on what Bricker calls the intrinsic conception of distance.

> If a space is maximally reinforced by direct ties of distance, then a distance relation such as "... being twenty feet from ..." is intrinsic to the points that stand in it. Whether or not the relation holds depends solely upon the intrinsic nature of the two points and *on the composite of the points*. Bricker (1993:1). My italics.

In a sympathetic account of this argument, Dainton describes what seems to be the same view, as follows:

> The distance between P and Q (and any other two points) depends solely on the properties of the points in question ... the distance between these points depends on nothing but the points themselves and *how they are directly related to each other*. This direct intrinsic difference is independent of the space within which P and Q are embedded. Dainton (2010:157-8). My italics.

---

[4]Work in the topology of spacetimes does allow for the removal of points and holes in the discussion of boundaries and singularities. How this fits with maintaining the metric distances is not clear to me (Earman (1989:Chapter 8 §3) Hawking and Ellis (1973:Chapter 8)).

[5]I owe the second thought to Peter Quigley.

What do the italicized phrases mean? Clearly, if the distance properties depend only on the points, then there must be a relevant property that each has intrinsically. But what is "the composite of the points" and what role does it play? Is it just the pair? One might say that the distance property is also intrinsic to the pair, although if that adds anything, there is no indication what it is. Dainton's "how they are directly related to each other" does not clarify the claim. How could the relation not depend on how they are related to each other? But how they are *directly* related is precisely the problem at issue. I don't understand whether or how the conjuncts imply a further condition. If the composite of the points or the manner of their direct relation is distinct from the properties intrinsic to the points, that appears to lay open the worrying possibility that P and Q might be 3 metres distant in virtue of their intrinsic qualities yet 4 metres apart in virtue of the composite of the points and the direct relation. Russell once retorted that he could provide an argument but not an understanding. Yet something more is surely needed to make sense of this. I press on in the belief that the view really is that distance is an internal relation, floated by properties intrinsic to the points.

2.7 Taking "intrinsic" seriously

According to the above, P has an intrinsic distance-property in respect of every other point. So does Q. That strongly reminds one of Leibniz, not in the style of his correspondence with Clarke (Alexander, 1956) but in the style of the Monadology (Leibniz, 1898). Monads have intrinsic properties that internally relate them to every other existent. Leibniz said much about what they are. It will not help to slip these intrinsic distance properties in as bare primitives: they are, at the very least, quantities that the real number *numbers*. To make sense of this we need to know, for one thing, what distance units and standards the number is tied to so as to know that it is not magnetic charge or gravitational potential. It can't just be called intrinsic distance – we have no glimpse of what that means yet.

There is a model for this, one that Leibniz took seriously. It sheds light on the kind of thing that a proponent of this view needs to tell us. Think of the properties intrinsic to a photograph as fixing the point in space and time of the camera when it took the photo. Then think of perception analogously: properties intrinsic to perceptions in the mind define for perceivers the point from which they view the spatiotemporal world. If these properties are truly intrinsic, then they do not need to look across a path to what is seen. Perception is all internal. (See Nerlich (1976, 1994:Chapter 1) for more detail.) Leibniz, as I understand him, seized on the general structure of this rather than its psychic aspect. The intrinsic distance theory needs some structural analogue of this.

Bricker's aim, like Leibniz's, is to take his points *out of space* as I have been describing it. The price is to follow Leibniz structurally, putting space inside points! Thus each point is some analogue of a Leibnizian monad, internally related to everything else by means of properties intrinsic to it and not by external relations such as paths. These properties have to be at least a bit like perceptual ones – structurally. I have no wish to foist a literal mentalism onto Bricker and Dainton. But just the structure itself makes their view of the world both abstract and strange. In particular, we lose the core principle of the concrete, that two things might have all their intrinsic properties in common yet be in different places or happen at different times. We are lumbered, instead, with the dubious Principle of the Identity of Indiscernibles. Leibniz saw this as a triumph for his metaphysical aim to confine all there is to substances and their qualities. The Monadology is a brilliant and consistent (as far as I know) intellectual structure but, for most modern minds, an implausible one. The garland of the Principle makes the victory hollow.

Further, Leibniz's claim (Alexander (op. cit: Fifth paper §5) "... one point of space does not absolutely differ in any respect from another point in space ..." is surely the standard conception of points, at least of points *in space*. Of course that is exactly why he wanted to be rid of them and make them differ solely in internal relations. How Bricker and Dainton would have us conceive of them is not yet clear.

Points are usually ruled to be immovable. But we must consider how it makes sense in this theory for objects to move. Let's suppose, then, that objects, like points, have their distances from each other in terms of pure metric space relations. Object pairs, then have real number distances as properties of the pair, and thus of each member of it as an internal relation (e.g "... is distant from Q") – if we grant the reflections that led us to distance as an internal relation between points. Suppose that just one object moves: then, for it, this amounts to a change in all its internal distance relations and it amounts to nothing more. Every other object changes one of its intrinsic relational properties, the one that distances it from Q. Nothing else happens. That is the sense of "travel" which this neo-Leibnizian theory of space and distance allows us. Space is a pure abstraction just as Leibniz concluded.

I don't claim this as a refutation of Bricker's view as he and Dainton defend it. I merely voice my disbelief in it and doubts how to understand it. I prefer the view of space as a concrete particular thing because I believe in the pervasiveness of paths in perception. We do see things so alike as to be perceptually indiscriminable yet different as to where they are or when they occur. I see no reason to think that such things must be, ultimately, discriminable in terms of unperceived intrinsic properties. Leibniz's space is a second order "well founded phenomenon" but its reality rests on intrinsic and somehow mind-like properties of things (in the sense loosely gestured at above). My metaphysics of spacetime makes it a first order, concrete thing pervasive in outer sense. That bit of ontology is not proposed as an a priori necessity but as a contingent truth, evidenced everywhere-and-when in our most familiar ways of experiencing the world. It's observational metaphysics.

There is a deep analogy between Leibniz (1898) and Kant (2007). This is not surprising given Kant's early devotion to, and continuing admiration of, Leibniz. Leibniz's problem was how to deal with external relations. Kant's problem was how to deal with the chasm that separates the objective from the subjective. Both arrived, in the end and for different reasons, at similar solutions. They may be summed up in an epigram that makes each of their theories a kind of Copernican Revolution: the mind is not in the world; rather the world is in the mind.

2.8 Substrata

"Substance" and "substrate" in Earman's definitions, are seriously misleading when applied to path realism. They suggest some connection with the bad old doctrine of substance as a "bare somewhat-in-which-properties-inhere:[6] substance is a featureless support or substrate for properties. Thus the label "substantivalism" for any spatial realism carries a hint of derision with it, whether or not that is intended. However, paths play no role as substrates. The phrase "... is separated from and connected to..." has a grammar and semantics like "...is chained to..."; that is, the grammar of external relations mediated by an entity independent of what it relates. For spatial relations among things, the entity is a path, and a particular path for particular relata. Among spatial parts, the entity is again a particular path and the whole complex a part of larger spatial particulars, i.e. regions, and finally the grand particular, space itself.

Paths are not substrates of *detachable* thing-to-thing spatial relations; there are no such relations for them to underlie. Paths are the particulars, the principal relaters, *directly* separating and connecting things to things as well as parts of space to parts of space. They are not abstractions of them or of anything else. The substrate or substance of a chain is a something (we know not what) arrived by

---

[6]'Inhere' presumably just so as to avoid the brutally crude 'adhere.' That is part and parcel of doctrine about substance and helps to convey a bad odour; it, too, taints substantivalism.

abstracting all the properties of the chain that are said to inhere in the substrate: the substrate neither exists without the chain nor the chain without it. A chain may occupy a path but no path is a substrate for a chain's properties. Remove the chain and you leave the path; paths need no chain properties; further, a chain may occupy different paths at different times but never has different substrates. Path realism is not a substance doctrine. Paths are not tainted with the ancient absurdities of substance. Realism is not substantivalism. Paths are concrete, particular but not material relaters. That's their ontic peculiarity. That's what realism delineates.

The usual definitions of substantivalism and relationism exemplified above implicitly endorse the form of dualism of spatial relations discussed in Chapter 1.5. They place it at the core of the traditional debate. They capture the crucial difference between the major traditional positions. However the definitions bear no relation to the arguments brought against the shift symmetries. The words "direct" "parasitic" and "substratum" don't feature in the critique of the shift argument whereas "dependent" and "independent" do. The quote from Leibniz that begins Chapter 1 may reveal an underlying reason for the difference. It is, again, his focus on the indiscernibity of *points* and on shifts as detaching things from, and re-attaching them to, indistinguishably different *points*. That casts the substantivalist as defending a sort of instance-by-instance dualism. For each particular relation of a thing to a thing, substantivalism cites another relation – the relation the *point-place* of the first thing has to the point-place of the second. It pays no heed to regions or their structures. This other relation is conceived in the definitions as standing to the first, the thing-to-thing relation, as a substrate stands to its attributes. Nothing like this goes on in the critique of the shift arguments: they fail because they do not recognise that thing-to-thing relations depend on the structure of space globally or region by region. They show that when things are shifted they cannot carry with them their old spatial relations to other things unless that is permitted – a rarity – in a new suitably structured region.

Realists who argue from backfiring as a principal ground of their claims are ill described as substantivalists. Of course they regard space as a substance in the weak, formal sense that it is a subject of attributes. That does not commit them to regarding the spatial relations that hold between point-places and are left behind in a shift, as pure substances without spatial properties. There are two kinds of *terms* that stand in spatial relations – things or points (places). There are not two kinds of spatial relaters, one underlying the other.[7]

---

[7]Brotherhood is a different relation from sisterhood only because the terms that stand in the relation are different. But the relator, the relation proper, i.e. siblinghood, is the same in each case.

The path that connects the places is the path that connects the things. Places don't have shadows of the relations to each other that the things occupying them have. Realists should energetically oppose substantivalism as these influential definitions portray it. They do not conceive of space or its parts as anything even remotely like substances in the traditional metaphysical sense.

I have laboured this theme since it is so difficult to rid realism of the taint of the absurd doctrine of substance, especially since it is suggested by the word "substantivalism." Realists continually find themselves labelled in unjustified and vaguely derisive terms.

2.9 Waddaya mean – concrete-immaterial?

I conceded earlier that unoccupied parts of space are not objects of perception even though we may trace them if they are paths. That is more a matter of usage standard in our world than it is of the ontic type of parts of space.

Suppose, in world in which space is Euclidean in the large, that there are football sized regions of space that have rather acute curvatures. We could directly see where they were, if nearby, since they would distort our vision in the way a heat haze can. We may think of experiencing this as rather like looking through and beyond the distorting football. This suggestion has a real analogue in observations such as the bending of light rays round the sun, visible in eclipses.

If you tried to push your hand into a non-Euclidean football, your hand would need to change its geometry so as to have a shape possible within that spatial region. You would feel that change as you pushed your hand in. There would be a resisting force. But the force is not exerted by the empty non-Euclidean football. *You* exert the force needed to distort your hand so as to get it into the football because the elastic (electromagnetic) forces that keep your hand in its equilibrium state in Euclidean regions will resist the change. Each particle in your hand must change its distance from all the rest. Your push must overcome the electromagnetic forces that would keep the hand in its Euclidean shape. *You don't push against the space and it does not push against you.* You push against forces internal to the hand. This idea has a real analogue in what can happen to things – what would happen to an astronaut, falling into black hole. There you would vividly experience the concrete spacetime curvature. It would kill you.

You would see the time-curvature of spacetime if you could see the gravitational red shift of a massive galaxy. This looks like, but is not the same as, the Doppler red shift of light from a receding star. Gravitational red shift is observable and has been observed, although not at domestic scales.

The examples help to show what I mean by saying that space and spacetime are concrete yet immaterial – the latter because space or spacetime exerts no force on anything in these cases nor are forces exerted on it. It plays no dynamical role and is not causal in that sense.

The topic is developed in more detail in Chapter 7.

2.10 The modesty of realism

The arguments so far defend a minimal spatial realism. They are drawn from observation and from the metaphysics and semantics of spatial relations. These cast paths as entities that mediate spatial relations among things, entities that are not of the same ontic type as the things of physics or recognised objects of perception. The properties of things that are terms in spatial relations shed no light on spatiality: the mediator, the relater does that. Space is the fusion or mereological sum of all paths. These grounds sustain little of the detail in terms of which we think about space.

The entrenched belief, and the semantics it recommends, entail not just that paths are independent of the properties of the things they separate and connect but also that the things they connect are largely independent of the nature of paths. Any kind of *spatial* thing at all can be supposed connected and separated by paths, without erosion of sense. That makes it meaningful to think of paths as separating and connecting parts of space as well as objects. That in turn makes *formal* sense of space empty of objects. It does not, and realists need not, either commit to or reject an independence of space from objects in any stronger sense. But positing a path connecting the number 3 to the number 4 remains senseless.

So far these arguments are best seen as metaphysical. They tell us little about paths, not even that paths are strictly one-dimensional or straight. They are parts of space considered in respect of length, ignoring thickness. This is vague but more precise ideas are mathematical not metaphysical. Much of our common understanding of paths is not a philosophical one and so not properly constrained by the sort of arguments on which our main topic depends. I sketch a few plausible empirical features to outline what I call *Modest Realism*.

In well-behaved spaces, an observed multitude of objects yields an observation that there is a multitude of paths only some of which are shortest (extremal). They separate and connect parts of space as well as things. Paths intersect in spatial parts (roughly points) and intersect surfaces (sums of paths). Some paths have paths as proper parts, perhaps separated by these objects. Perhaps all paths have paths as proper parts. Only here do we strike an issue about the adequacy for discourse of spatial relations just among things. These further steps look plausible once we take the first. They add no ontic types or conceptual load to ontology, but only more paths.

Rash Realism is the view that continuity and differentiability are among basic primitive properties of any spacetime. These postulates have a pragmatic motivation - there to legitimate the use of differential equations and calculus generally. It seems intuitive that space is infinitely divisible i. e. dense. That is not without its problems however. There is no *metaphysical* reason to accept these views of space, so none to accept points. No realism presupposes any particular global topology, nor any particular metric.

Further questions would take us far from our metaphysical concern with the spacetime of GR. The problems of continuity and even of denseness are difficult but important. We do not know whether these properties belong to space and time.

Calculus and differential equations, so dominant in physics, are part of Rash Realism for pragmatic reasons that may lapse with further developments in mathematics and physics. But an understanding of crucial topological concepts is also at stake. Consider the dimensionality of space and time. We take a space to have n dimensions if regions may be separated from each other by a finite number of boundaries of n-1 dimension. Thus a closed curve will cut the plane[8] into two parts, neither reachable from the other without crossing the curve. That shows the surface to be 2-dimensional. But a curve is 1-dimensional only if it has no width, no surface area for its cross section and so on. We do not observe that there are such things either as material objects or as spacetime paths.

Approximation works. A closed material thread approximates a geometric path-boundary on a tabletop. If it has length and is closed we can neglect the magnitude of its cross section. Its length and breadth, not its thickness, illustrate the dimensions of the table's surface. The table's surface itself is the uppermost part of the table considered as extended in two directions without our specifying a depth. So if, in practice, we can fudge the elegant account of dimension found in topology by ignoring aspects of material things, we may do the same in geometry itself. For our purposes, we need not insist that paths are 1-dimensional in the strict sense.

I say no more about these issues, enticing though they are. In the end (chapter 7), the fundamental posit of GR, i.e. what is presupposed in formulating the fundamental field equation, is just the modest structure described here.

---

[8]It will do the same for the spherical surface. But some closed curves do not divide the toral surface into distinct regions. To describe which curves don't is a fundamental way to define the topology of the torus (donut surface).

2.11 Reflexive spatial relations

Nothing in the entrenched beliefs forbids spatial relations to be reflexive. Things may be spatially separated from and connected to *themselves*.[9] Reflexive spatial relations can be proper relationist relations since their terms refer to observables. Things may be self-connected and self-separated by many different paths including shortest (extremal) ones. The North Pole is at a longitudinal distance from, and connected to, itself via infinitely many equal yet distinct longitudinal geodesics. Similarly, there are possible worlds in which only one observable thing exists and is separated from and connected to itself by infinitely many distinct geodesics. Consider, for simplicity, the example as set in S_2. Let the object's place be regarded as a pole of the space. Then an equatorial path separates the distinct points where it intersects each reflexive geodesic. Without some such structure nothing constitutes the distinctness of the different reflexive relations so as to constitute the geometry of S_2. The relationist has no way to explicate how there can be distinct reflexive relations in such an example.

In Max Black's well-known counter example to the Identity of Indiscernibles,[10] it is sometimes assumed that, because an iron sphere is at a spatial distance from an indiscernibly different iron sphere, that they are indeed two, since the relation is assumed irreflexive. But it need not be. What makes them two is not just that they are in the same space, path-separated and connected. In the spherical space, S_3, there could be just one object self-separated-and-connected by a path that is a circumference of the whole space. *That would not breach any relationist principle.* Clearly, there would be many such paths. Thus more than one or two distinct spatial relations – paths – are needed to give space the global topology of S_3, rather than E_3. The physical nature of iron spheres cannot yield anything like that.

2.12 Talking about paths

A modest realism merely admits space to ontology as a first order entity, conceptually unique and primitive.

Although they pervade perception, paths are certainly a discursive problem. It is easy to remind ourselves of what we see. How to *talk* about paths is less straightforward, especially just how to de-

---

[9] Of course metrically non-privileged paths provide for that. There are also many non-privileged (intrinsically curved) paths linking the Pole to itself.

[10] Black (1957:153-164). The example, put neutrally as regards identity or difference is this: suppose a world in which a sphere is one kilometre from a sphere and no property of any sphere is different from any property of any sphere.

scribe their ontic type. But if the argument so far is sound, it may be useful to hazard a preliminary description: paths in space (spacetime) are concrete particulars, immediate in perception, independent of the properties of matter, formally objects of reference and bearers of properties. They are concrete but not material. Indeed, they are of the essence of the concrete. To be concrete is just to be distinct despite indiscernibility in intrinsic properties or relations dependent on them. It is spatiotemporal relations, mediated ones, that distinguish concrete things. It is precisely this sense of concreteness that Leibniz Equivalence seeks to undermine.

The chapter's main conclusion is this: a metaphysical theory of spatiality entails that the properties of space and time are not founded on objects or their properties, but are path-mediated. Paths are not explicated though the nature of their occupants or terminants. Space is their fusion.

3 On the Sovereign Independence of Spacetime

> Chapter 3 moves from pre-1905 physics to spacetime along what Minkowski called "a purely mathematical [i.e. geometrical] line of thought." It's begins from the symmetries of classical mechanics, is independent of electromagnetism and draws only on spatiotemporal features of classical mechanics. It ends with the structure of spacetime as providing the basis for the Lorentz transformations. The transformations merely reveal the basis. The independence of spacetime from dynamical theories generally is shown. Thus field theories in general, and also mechanics and electromagnetism (and light) must, and do, conform to the structure of spacetime. It is sovereign in that respect.

3.1 Introduction

SR is not a branch of electromagnetism: it is not founded on light's having a constant, limiting speed. If the theory is true, it is independent of all matter theories. Rather, it springs from general and familiar symmetries of space and time: these impose the form of the Lorentz transformation on every matter theory, independently of any dynamical content it may have.

Minkowski first developed this revision of SR in §1 of his path-breaking paper of 1908.[1] That part of his message has never been near the centre of attention. For some, it entered the basic folklore of SR – e.g. for more mathematically minded physicists. Many more know nothing of it – especially the majority of philosophers, includ-

---

[1](Minkowski, 1908). In this chapter numbers in brackets refer to pages in this paper.

ing perhaps a majority of philosophers of science. For these, SR rests on the principle that nothing outstrips light, made vivid by the sheer brilliance of Einstein's operationalist analysis of light-synchrony and simultaneity. That is how most people learn the theory. It has overshadowed the geometrical approach.[2]

But Minkowski arrived at the Lorentz metric, at spacetime and, therefore, SR "along a purely mathematical line of thought" – that is, along a purely geometrical path from the symmetries of real, i.e. physical, space and time to the Lorentz transformation and the metric of spacetime (75). What earned first place in Minkowski's lecture is the mathematical purity of the journey rather than where it began. Einstein's revolution in physics could have been gained by bold mathematical conjecture based on familiar, empirical symmetries of space and time. Minkowski did not just improve on Einstein's theory: he gives it quite new foundations. It needs neither light, nor electromagnetism nor any other matter theory to sustain it.

The famous four-sentence preamble to the lecture acknowledges that the new views of space and time "have sprung from the soil of experimental physics, and therein lies their strength" (75). But Minkowski then goes straight to the mathematical approach. He concludes "... mathematics, though it can now display only stair-case wit, has the satisfaction of being wise after the event" (79).

Thus, in SR, spacetime itself lies at the foundation of matter theories, imposing its constraints on each one. Its structure yields the broad picture of the world that has shaped physics since 1905. Spacetime is sovereign and independent.

I shall amplify and simplify Minkowski's rather informal discussion so as to track philosophical interests. It is deeper, more intuitive and generally accessible than the more analytic and algebraic deductions that followed in 1910 (see §3). Its metaphysical, ontological importance is my main theme.[3]

---

[2] In the vast philosophical literature on the conventionality of simultaneity, the shade is deep indeed. That no signal outstrips light is a dominant theme, I believe a misleading one, pervasive through that work, from Reichenbach onwards. It did much to lend the discussion a spurious epistemological flavour. Conventionalism was the main focus for philosophy of relativity in the 60s and 70s. At the very least, the issue looks different within Minkowski's picture. In general, the frequency of use of "light cone" "light line" etc. far outstrips "null cone" and "null line."

[3] For an informal discussion with a reasonably accessible preface preceding a complete mathematical treatment see Cacciatori et al. (2008); Dyson (1972:635-72) is helpful and informal. Bacry and Levy-leBlond (1968:1605-1614) is a succinct and somewhat different analysis, accessible to those familiar with Lie groups.

3.2 Minkowski's "purely mathematical line of thought"

Geometry was widely viewed as an empirical study among the mathematicians of Minkowski's time. This was no less true in Göttingen.[4] Hilbert's interest in axiomatics did not lead him to think of geometry only in terms of uninterpreted formal systems. It was part of physics. Minkowski shared that view. Yet he concluded that his geometry for spacetime had to impose its structure on the foundations of any foreseeable theory in physics.[5] The line of thought is more mathematical than physical. However, it is not a line of thought in pure mathematics.

By "purely mathematical line of thought" Minkowski clearly meant, in particular, a line that does not draw on the physics of light as the source of his derivation. It draws on the symmetries that sustain the neo-Newtonian version of the relativity of motion and the uniformity – the translation symmetry – of time. Behind each of these there lie centuries of observation, theoretic speculation, blunders and confusions about empirical problems that painfully led theories of motion to the neo-Newtonian version its relativity. It states that uniform motion (constant velocity) is a symmetrical relation among bodies so long as they are at rest in some inertial frame. It may also hold among these frames themselves. That is not a priori. A more radical version, that it is an epistemologically necessary truth that motion is a symmetric relation among observable bodies, was long a dogma among relativists about motion. But that thesis came to be wholly abandoned in GR. Any coordinate system at all may be taken as defining a rest system without regard to whether any body is at rest in it. Minkowski's version of these symmetries is contained in "the accepted mechanics of the present day." They have "sprung from the soil of experimental physics, and therein lies their strength"[6]

Minkowski, (76)[7] begins by noting two groups of transformations that leave Newton's laws invariant: changes of position and orientation and changes in uniform velocity. The first is the standard Euclidean group of continuous spatial transformations – translations, reflections

---

[4] See Corry (2004); Walter (2008).

[5] For a discussion of the transition from Einstein to Minkowski see Norton (1993:791-858); Corry (1997:281-314); Walter (1999:45-86) and Walter (2011).

[6] Op cit.: 75. Here "the present day" means before 1905. It was the physics that Einstein and Minkowski revised. The four-vector approach to various key mathematical concepts was already developed by Poincaré. But it was mainly Poincaré's conventionalism that led him to discount spacetime as a significant player in his theory. For his role in this see (Walter, 2011) and Whittaker's (1910:vol II: 64) more dismissive mention. I am indebted to Scott Walter in this paragraph.

[7] The next section makes clear that Minkowski needed just the symmetries of Euclidean space and the reversibility of velocities.

and rotations. This means that the laws of physics are the same (invariant) everywhere and in any orientation. The second is the Galilean group of transformations that take us from one inertial frame of reference to another. "... the two groups, side by side, lead their lives entirely apart (76)". He aimed to unite them.

Minkowski reminds us that no one observes times except at places or places except at times (76). Thus, space and time, although deeply different are also deeply connected, a connection that neither group exploits. The two groups are incongruous in that only the first contains an orthogonal rotation and then only of the spatial coordinates. The second allows replacement of x, y and z with constant speeds in these spatial directions (79). Time enters into the Galilean group via this relative speed of frames of reference, and so appears in the transformation of x, y and z coordinates. Yet the time coordinate itself is not transformed: t remains t. This leads the Galilean group to be "treated with disdain" (loc. cit.) presumably because it thereby raises the embarrassing problem of absolute motion. Neither group envisaged nor determined any orientation of the t axis to the spatial ones.

3.3 The form of a new metric

Connecting the two in a satisfactorily unified group is surprisingly simple and rather conservative. It simply invents a new metric, as closely modelled on the Euclidean one as is possible. It is the sort of free creation of the human mind that Einstein so admired. Minkowski describes it as "fancy-free" later in this section, in suggesting that a mathematician could well have constructed the theory in this way before Einstein's approach through light-signalling. I have found in discussion that people who are familiar with the metric sometimes do not fully understand why it takes its simple form. I shall set that out in detail, elementary though it is.

An obvious conservative guess is that the t−coordinate should be orthogonal to the spatial ones just as they are to each other. The unified structure of the three spatial dimensions rests on Pythagoras's theorem. The line element, i.e. the metric, of space, i.e. the familiar differential equation that gives the distance between two nearby points in terms of their coordinate differentials, is:

$$ds^2 = dx^2 + dy^2 + dz^2$$

So dt should be included in the equation for the spacetime interval in a somehow similar way.

There are two provisos.

First, to tie space to time so as to preserve their deep difference within the unity requires different *signs* for dt and for the space differentials (their sum). This will be discussed further in following paragraphs. A unified metrical structure can then define the spacetime interval ds between two nearby events that bound the spacetime element.

Second, because spatial units and temporal ones differ, the equation needs a parameter, c, in units of space and time. For instance if measures along the x and y axes were given in metres but measures along the z axis given in yards then the spatial metric would need a coefficient for dz to record metres per yard. Then it would read:

$$ds^2 = dx^2 + dy^2 + (0.8)^2 dz^2$$

That gives the distance differential in metrically uniform and meaningful way. So the new parameter in the spacetime metric must play at least the role that (0.8) plays for metres and yards. It must tie spatial and temporal measures together. Thus c will have the units of a speed – metres per second, light years per year – depending on the conventional choice of units.

These provisos lead immediately to our goal:

$$ds^2 = c^2 dt^2 - dx^2 - dy^2 - dz^2$$

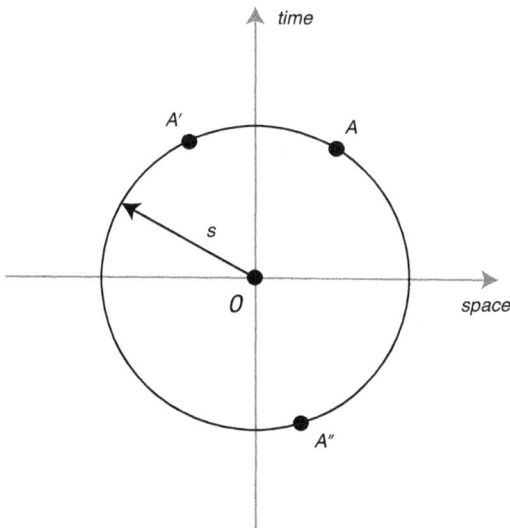

Figure 3.1: A Euclidean metric that does not distinguish the t axis by its sign in the metric does not rule out a frame of reference in which A'' is before O.

The metric of a space tells us everything about the structure of a space in a very small region round a point. Since it is a long entrenched part of physics before 1905 that space has the same structure at every point, the metric, including c must be everywhere the same. The equally long entrenched view that time has constant structure – the laws of physics don't change – means that c does not change. It is a universal constant.

If new transformations were to rotate space and time together so that the signs in the metric are all positive, that would preserve the connection between space and time but erase the difference. The modest step of giving minus signs either to the x, y and z squared differentials, or to the t differential registers the difference. It also prevents any transformation that would rotate a time axis into a spatial one.[8] Let **A**, **A**′ and **A**″ in figure 3.1 represent events at the spacetime interval **s** from **O**. The first two occur later than **O**, the other earlier. We may choose coordinates so that **A** lies in the positive direction from **O**. We can rotate the coordinates so that **A**′ lies at interval **s** on the positive direction of the new t coordinate axis. But we can equally well rotate them so that **A**″ lies on the positive direction of the new t axis. The transformations allowed by the metric do not conserve the time ordering of events even when both lie on the same timelike line. In short the metric does not give t a direction in the spacetime.

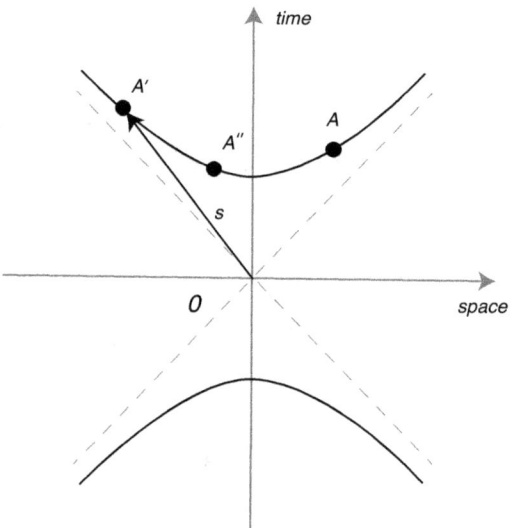

Figure 3.2: In 2D, the signatures $(+, -)$ and $(-, +)$ both give a hyperbola as the locus of points a fixed interval from O. Events later than e_1 are all plotted on the upper lobe, earlier ones on the lower lobe. The null cone is dotted. The lobes of spacelike separation are not shown.

---

[8]See Cox and Forshaw (2009:74-88).

If the difference between time and space is respected by a different sign for the t variable then the metric is very different. The locus of points at a spacetime interval from **O** is a hyperbola. At each point it separates the orientation of timelike lines from spacelike ones by the null cone boundaries within which lie all events at timelike intervals from **O**. The positive directions of the time axes of all coordinates are separated from the negative directions. All events preceded by **O** remain preceded by it in all frames of reference. As figure 3.2 illustrates, that time order is invariant under the Lorentz group of transformations defined by that signature of the metric.

The metric does not pick out which direction in time is past and which is future. Which half of the axis is numbered with positive coordinates and which with negative ones is a conventional choice as is the habit of taking the upper direction of the null cones on the page as indicating the future.

3.4 The value of c

So far only the form of the metric has been explained. Nothing yet determines the value of c although figure 3.2 assumes that it will be finite – more on this later. It must be found by observation, but not necessarily observations on light. The value must lie in the interval $0 \leq c \leq \infty$. Values within the limits are "mathematically more intelligible" (79) than either limit is, even though the upper limit defines the familiar Galilean group. A limit of zero speed is unbearably queer.[9] Infinite speed is queerer still: it is $dt/0$ and division by 0 makes no sense. Further, along its line of motion the thing is at each point simultaneously and must overlap itself! A finite, invariant limiting speed forbids velocities, as ordinarily conceived, to add arithmetically. This certainly offends one intuition but rescues others from difficulty: spatiotemporal continuity, and thereby both identity and causality make better sense.[10] Minkowski doesn't itemise these metaphysical grounds in seeing a finite speed as mathematically the more intelligible. Doubtless he recognised them. Thus, requirements of sense and elegance lead to a finite constant, c, for the Lorentz group, and so to the structure of Minkowski spacetime. While transformations leave physical laws the same in each frame, they no longer leave the properties of things invariant.

Group theory has general objective criteria for simplicity. The

---

[9]This defines the Carroll group. "A slow sort of country," said the Queen, "Now, here, you see, it takes all the running you can do, to keep in the same place." Carroll (1871:109).

[10]Lucas and Hodgson (1990:§1.1).

Lorentz group is semi-simple; the classical Galilean group is not.[11]

The upshot is a full rotational transformation group: change of inertial frame (boost) rotates all coordinate axes (through a pseudo-angle for time). A boost is an angular, geometrical, transformation. A thing (particle) is a material curve in spacetime, a worldline if it is straight.[12]

Here are Minkowski's own words on how an audacious mathematician might have been the first to find spacetime and SR.

> Group G_c in the limit when $c = \infty$, that is the group G_∞, becomes no other than that complete group which is appropriate to Newtonian mechanics. This being so, and since G_c is mathematically more intelligible than G_∞, it looks as though the thought might have struck some mathematician, fancy-free, that after all, as a matter of fact, natural phenomena do not possess an invariance with the group G_∞, but rather with the group G_c, c being finite and determinate, but in ordinary units of measure, *extremely great*. Such a premonition would have been an extraordinary triumph for pure mathematics. ... I shall state at once the value of c with which we shall be dealing. It is the velocity of the propagation of light in empty space.
>
> Minkowski (79) [Original italics]

In an earlier lecture, in 1907, he spoke of "a complete change in our ideas of space and time" and went on:

---

[11] Dyson, op.cit. : 642.

[12] Dyson (op. cit.) suggests, and Cacciatori et al. (op. cit.) agree, that this line of thought might have been continued past the point where Minkowki left it while remaining purely mathematical. The Lorentz transformations operate at a spacetime point and are homogeneous in the point. The complete group that captures all the symmetries of Minkowski spacetime must include spatial rotations and the group of translations T_4. This defines the Poincaré group. But the inclusion of T_4 makes this group inhomogeneous (Cacciatori et al. (op. cit.) §1.) Mathematically a simple and obvious terminus is at the form of de Sitter spacetime. It contains a further constant, the cosmological constant, λ, its value again to be found observationally. λ did not figure in Einstein's original field equations because he wanted Minkowski spacetime as the mass-free solution of the field equations (Rindler (1977:§14.5)). So, mathematical simplicity yields a de Sitter spacetime that, even though it is matter free, not merely expands but accelerates its expansion (or decelerates it or alternates expansion and contraction). Matter would perturb its structure merely locally. Recent observational evidence suggests that this is close to the structure the universe has (Clifton and Ferreira, 2009). Many physicists who fully accept Minkowski's conclusion, pause at the further move to de Sitter spacetime: it lacks the same motive and inevitability. I will not pursue the matter but refer readers to §1 of Cacciatori et al. op. cit.). For a robust defence of the crucial importance of Minkowski spacetime see Petkov (2010).

The physicists must now to some extent invent these comments anew, laboriously carving a path for themselves through a jungle of obscurities, while very close by the mathematicians highway, excellently laid out long ago, comfortably leads onwards.[13]

Clearly, the line of thought was presented in 1908 as speculative – fancy-free, not deductive. Neither its premises nor its conclusions are a priori or necessarily true. It has to match the world.

3.5 A deductive approach

Newton's title "Corollary V" for his discussion of the motion of relative spaces implies, essentially, that his Laws of Motion entail Galilean relativity. It must have seemed obvious that relative motions will inevitably retain, as invariants, all properties of things that could consistently be taken as invariant. That delivers the Galilean transformations. Yet Newton made no serious attempt to derive them. We now know that the conclusion does not follow. Einstein gave the decisive counterexample in 1905. He was motivated by the belief that a satisfactory understanding of Maxwell's electrodynamics required the speed of light to be a finite constant, invariant under change of inertial frame of reference. Yet the Lorentz transformation was never confined to electromagnetism. It led at once to a Lorentz invariant revision of the laws of mechanics. From the start, the transformations were understood as a formal demand on any matter theory in physics. They are at the foundation of all.

A question naturally arises: what constraints on transformations does the Relativity of Motion, just by itself, demand? This is the better-known, more deductive approach (Ignatowski, 1910). Its starting point is closer to the issue of the addition of velocities. It demands the same deep and familiar symmetries of space and time exploited by Minkowski and long taken as a priori necessities in Euclidean geometry. More explicitly, assume the homogeneity of space and time (laws don't change over time), and isotropy of space.[14] Despite Minkowski's saying he begins "with the accepted mechanics of the day" i.e. classical mechanics, his conclusion actually obliges us to revise it. It is really from these geometric premises, underpinning the relativity of motion,

---

[13]Minkowski (1915:372).

[14]This is usually said. But you also need a kinematic isotropy: if F_2 has velocity **v** relative to F_1, then F_1 has −**v** relative to F_2. See Torretti (1983:79-80). Feigenbaum (2008) notes that standard deductions assume for simplicity that the frames of reference are in standard configuration, thus sidestepping the need for a Wigner rotation (aka Thomas precession) if the relative motions are not all in the same plane. The full Lorentz group is not commutative.

rather than classical mechanics in full, that both arguments begin. It yields the form of the Lorentz transformations with an undetermined constant, c, just as in Minkowski's more intuitive approach.

Work in this style began with Ignatowski as early as 1910. There is now a considerable physics literature[15] proposing more or less minor variants on similar routes to the same conclusion. Again, appeal is made neither to electromagnetism and light nor to assumptions about simultaneity.

For an excellent more physics orientated discussion of spacetime see Stein (1968).

3.6 Field theories and SR

A field is a continuum, not a complex of spatial relations among things. For classical fields, variable quantities, usually the values of some matter theory such as electromagnetism, are assigned continuously and smoothly, to points of spacetime.

Spacetime underlies the field and not vice versa (See §10.3.3). Fields presuppose continuity and differentiability. They are imposed on relations of path separation and connection to yield points at which field values hold. We have no understanding how matter theories or field values could provide any of this structure. Every field presupposes it, each in the same way. A field structure of paths and points is independent of how the field is sourced physically, or of how complex its values (vectors, tensors) may be at points. Electromagnetic field values do not spread spacetime; spacetime spreads the field values. Our current concept of (non-abstract) field makes spacetime prior.

Quantum physics may change that. One gauge-theoretic approach to GR would reduce spacetime curvature to a spin-2 particle (graviton) gauge-field. So far, this is more a hope of quantum gravity than an achievement.[16] Various attempts to place GR on a non-geometric (e.g. a bimetric)[17] footing are current but none has yet succeeded. Richer geometric structure in string-theory would supersede the present approach. But, in standard GR, e.g. as we find it in the mainstream chapters of *Gravitation*,[18] spacetime, much as Minkowski invented it,

---

[15]See, e.g. Mermin (1984); Sen (1994:157); Feigenbaum, op. cit. For treatments of the deduction by philosophers see Arthur (2007), Brown (2005:105-109), Lange forthcoming, Torretti, op. cit. §3.7 and §4.1 − 4.3 on Minkowski Spacetime, where the geometric structure is derived and described in detail, in a purely mathematical way. Curiously, Pauli (1958:11), mentions Ignatowski's approach, says that it should have been attainable simply by group considerations, but does not mention Minkowski.

[16]See Mills (1989:493-507); 'tHooft (1980).

[17]Brown op. cit. §9.5.2 claims that the bimetric solutions are not geometric, although he notes that Bekenstein (2004), refers to the geometric part of the action.

[18]Misner et al. (1973)

is the principal entity. I am arguing for a realism that understands "spacetime" in GR as directly referential without a need to reparse it.

3.7 Electromagnetism

The impression may well remain that electromagnetism plays a dominant role in fixing the finitude and the specific value of the fundamental constant c. However, for many theories, the observed value of a principal constant is not rooted in their structure. Only observation yields the values; as far as current theory goes, they are brute. This is a strong claim, but it is directly acknowledged in the many "fine tuning" arguments afloat. These note that many fundamental constants are improbably finely adjusted to each other in ways not demanded by the fundamentals of the theory. Without that tuning, life as we know it – indeed any life at all – would be impossible. Thus Paul Davies:

> In the present state of our knowledge the 20 odd parameters that appear in the standard model of particle physics seem to just be completely free, they're not determined by any underlying theory. But what is clear is that if some of them have values even a little bit different from those that they do, then there could be no life in the universe.[19]

So constants could be re-valued without insult to the structure of the theories that give rise to them. A derivation of dimensional constants, such as c, looks very unlikely. That the fine-tuning is improbable may point to our limited knowledge. We need not suppose that constants are somehow arranged to give humanity a ride.

This view of c in particular, is tacitly endorsed in many good popular versions of SR. For instance, Gamow's well-known early story[20] sets c at 10 miles per hour. This rewrites the covariant consequences of the theory at domestic scales without changing the structure of electromagnetism or mechanics in SR. For instance, c may still be defined as the ratio of the electromagnetic to the electrostatic unit of electricity. All the features of Gamow's story fit the general physics perfectly – it is electromagnetism, but slowed down.

Light moves at the constant invariant maximum speed because it is a zero mass particle; light doesn't determine what that speed is.

---

[19] Davies et al. (2006). See also Rees (1999).
[20] Gamow (1957).

3.8 Mechanics

Newton's first law underpins the relativity of motion in its standard form: the laws of physics are invariant under transformation to inertial frames of reference (privileged frames). The law is:

> Every body persists in its state of rest or of moving uniformly straight ahead, except insofar as it is compelled to change its state by forces impressed.[21]

Already this broadly hints that not all of mechanics is matter theory. Remarkably, the law says nothing at all about the properties or causal powers of any body to which it might apply. Remarkably, too, we have good classical reason to think that no body ever escapes the (gravitational) causal net or persists in its state of rest or uniform motion (although there might be some bodies on which the resultant of forces is zero, briefly or not). The law is not about bodies; it is about trajectories, spatiotemporal entities, not about what might occupy them. The same is true for SR.

Thus mechanics does not imply or suggest that the structures of space, time and spacetime derive from the physics of matter.[22]

3.9 Metaphysics, not physics

This is an argument in metaphysics, not physics or mathematics. It's about spacetime, a concept that plays a major role in GR. There it is a fundamental, highly articulated entity rich in explanatory consequences within the most powerful, elegant and intelligible theory ever to grace physics. GR may fail. Till then, the ground-floor place of spacetime in present ontology, its clarity and its central role in one superb theory, has to be of fundamental metaphysical interest, well worth positive exploration rather than a reflex impulse to expel or materialise it.

GR was the first theory of physics to be formulated presupposing only minimal structure for its background space. This is the differentiable manifold. Its weak structure is, and is required to be, hospitable to a wide range of geometrically different models. Only separation/connection, continuity and smoothness are given in the manifold. Other geometric and matter (dynamic) properties are encoded in tensors (complexes of vectors). Thus the most basic properties of space and time, separation and connection of tensor locations, are clearly

---
[21] Newton (1999:416).

[22] For an interesting discussion of the prominence of spacetime metric in the foundations of GR see Lehmkuhl (2010).

primitive postulates of GR. Smoothness and continuity get in, pragmatically, to justify the use of points and differential equations. There is no understanding how to encode these deepest features of spacetime from any tensor field envisaged in GR, else that would surely have been done. They are accessed in familiar experience.

Last, the symmetries to which Minkowski appealed are global in spacetime. The spacetimes envisaged in the full range of GR models have those symmetries only locally. But, significantly, they do hold in the tangent spaces common to each point in any model.

4 On the Benefits of Four-Thought[1]

> Chapter 4 presents a standard derivation of the most famous equation $e = mc^2$ stressing throughout that it depends almost entirely on the geometry of spacetime and especially the analysis of 4 vectors. Mass is the only relevant non-geometrical primitive. Thus spacetime structure plays a strong and prominent role in SR physics.

4.1 Introduction

In the context of Minkowski spacetime SR mechanics is recast in 4-vector (and tensor) form. 4-vectors especially highlight the role of spacetime geometry. It is too seldom stressed how important and simplifying a role vector geometry plays. Much that seems forbidding and complex in 1905 SR becomes direct and inevitable in Minkowski's theory. That is the focus of this chapter.

4.2 Representing worldlines

Minkowski's SR is the study of worldlines – the trajectories of objects in spacetime Minkowski (1908:76). A worldline is any occupied timelike path. If the path is straight, it represents a thing at rest or in uniform motion in some frame. If curved, it represents acceleration. At each point a tangent 4-vector represents its 4-velocity then and there. It tells us the direction in spacetime of the worldline at the point. That the representation succeeds is not guaranteed by pure

---

[1] Good accounts of the maths and physics of this topic may be found in Taylor and Wheeler (1992:Chapter 7) and Lange (2002:Chapter 8). The latter is written by a philosopher. My aim is to clarify and stress the dominant role of geometry but add nothing to the physics of the issue. A simpler account may be found in Cox and Forshaw (2009) Chapter 5 especially pp.122-39.

geometry: it is a postulate that the basic mechanics of things in spacetime is best developed in 4-vector geometry. Then we can test whether the representation works in the real world.

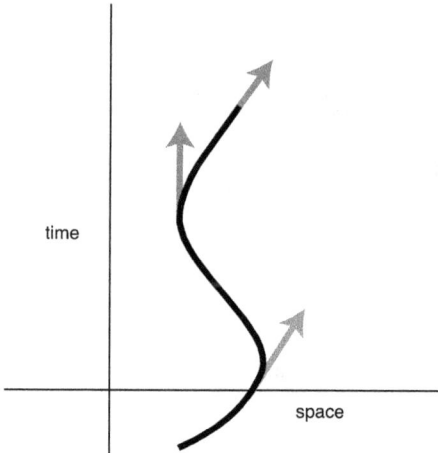

Figure 4.1: Velocity vectors tangent to a curved worldline

Thus there is a velocity 4-vector, tangent to each worldline, one at each point (event) on it. A velocity vector has a magnitude given by an analogue of the particle's speed at that event. But it is not the same as the 3-quantity ds/dt that gives the particle's speed in a reference frame. The velocity vector represents an absolute property of the absolute worldine at the point to which it is tangent. For nearby points on a worldline the velocity vector's magnitude is given by the spacetime interval traversed, divided by the proper time $d\tau$ taken to traverse it, as measured by a clock on the worldline. The interval traversed is the proper time taken, $d\tau$, multiplied by the constant speed-conversion factor, c. This is then divided by the proper elapsed time, $d\tau$. That gives, as a spacetime *analogue* of speed, $cd\tau/d\tau = c^2$.[2] The magnitude of each velocity vector is, therefore, the same as that of every other, c. It does not measure a speed but indicates only a direction in spacetime.

---

[2]The constant c represents a speed in units of space and time. Units may be chosen so that the constant is 1. Choose lightyear as a unit of space and year as a unit of time: then light travels one light year per year. More often, metres are chosen as units of space and the name is also given to the unit of time. A metre of space is set equal to $300,000,000$ metres of time. Thus the finite constant speed $c = 1$ metre/metre. Since 1 is an identity in multiplication c and its square are frequently omitted from equations so as to display clearly the main content of them. That will be done frequently in this book. However even when the constant is unity it is good to remind oneself that the idea of an invariant finite constant is pervasive in these equations.

Minkowski wished to describe his theory as *the postulate of the absolute world* rather than the postulate of relative motion. It is the worldline, an absolute, that is the new focus. Velocity 4-vectors differ significantly (and absolutely) from each other only in their *directions* in spacetime.

This representation of a thing's possibly varying motion by a worldline and its possibly varying tangent velocity vectors at different time points is purely geometric and uniquely apt: once we choose to represent mechanics in spacetime, this way of doing it is geometrically inevitable. Note that no property of the thing (particle) plays any part in it.

The momentum and energy of a particle at a time are represented by another, single, 4-vector, the energy-momentum vector It is simply a scalar product of the particle mass with the velocity vector – a representation that is, again, inevitable in the spacetime context. Momentum and energy make separate sense only in terms of the different *components* of this vector in some frame of reference. Relativistic momentum for motion along an x axis is thus $m\,dx/d\tau$, relating the vector to a frame. It differs from Newtonian space and time momentum, $m\,dx/dt$, relating a particle's space-and-time motion relative to the frame.

4.3 Representing mass, momentum and energy: $e = mc^2$

In spacetime, mass is a *non*-geometric parameter of the particle, its proper mass – "proper" in being intrinsic to the thing, the thing's *own* mass. Mass, here, is strictly inertia, i.e. resistance to acceleration.[3] Otherwise we may regard it as primitive; its nature plays no explanatory role in the geometric development of conservation and "mass-energy equivalence." Proper mass is an invariant of the Lorentz transformation.

---

[3]*Relativistic* mass is mass relative to a frame, a function of proper mass and speed (given by the components of the velocity vector) relative to a frame. This gives rise to the idea of *rest* mass – mass relative to its rest frame. But proper mass is not a relation of thing to frame; it characterizes things intrinsically. Relativistic mass vanishes from more recent accounts of relativity. In the expression of relativistic mass, proper mass may appear, conventionally written as m_0 – mass at 0 relative speed or rest mass. But proper mass is the dominant mass concept.

Mass is unlikely to remain a primitive scalar. There are theories about it in quantum mechanics (the Higgs particle if it exists: see Cox and Forshaw (2009). There is electromagnetic mass, about which there is little agreement. (See Feynman (1965), vol. I, §28 − 3f; Petkov (2009) Chapter 10 gives an illuminating discussion of the theory; Lyle (2010a) provides an extended account). But a structure for mass seems unlikely to affect the argument here.

64

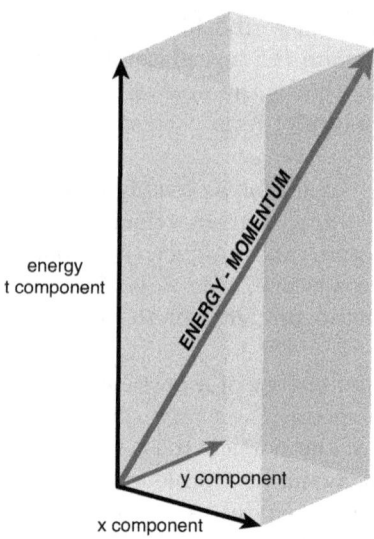

Figure 4.2: Energy momentum 4-vector shown with components.

In the rest frame of a particle, the space components of its energy-momentum vector vanish. There is no kinetic energy either. Yet the time component of the vector cannot vanish since it encodes the magnitude of the velocity vector times the mass of the particle. That introduces something new. The time component $m\,dt/d\tau$ is deemed the total *energy* of the particle. When it is at rest and the kinetic energy is 0 the *total* energy is given by the magnitude of the vector's time component *and is thus identical to the mass of the particle*. If we restore the constant c^2 to remind us of both the conventional ratio between seconds and metres, and the limiting constant invariant speed, then we naturally write this as the most famous equation in physics:

$$e = mc^2$$

It is crucial to realise that this equation involves just the rest energy of the particle. Kinetic energy is no part of it. The equation has simply fallen out of the inevitable way of formulating spacetime vectors.

But how simple and obvious is it that the component $m\,dt/d\tau$ is the total *energy* of the particle? When we are not in the rest frame, the kinetic energy must be represented within the time component since there is no other way to represent it. But in the rest frame, every material thing still has potential energy of some kind – heat, stress, chemical, binding energy and so on. In classical physics, where momentum is the integral of kinetic energy, any constant potential energy may be added to the kinetic energy of a particle without offence

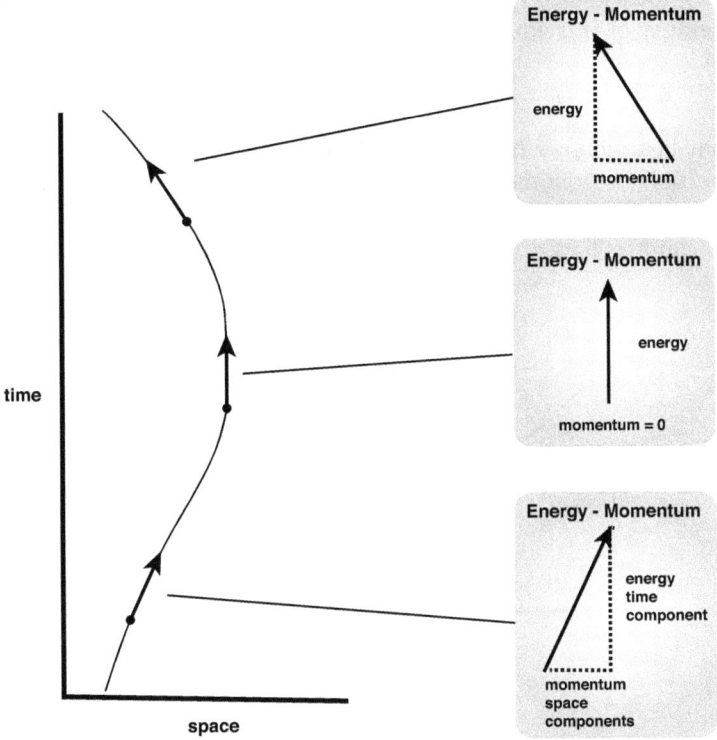

Figure 4.3: Energy-momentum vectors at different times of an object in mainly accelerated motion. Its components depend on the frame of reference portrayed here. Components vary as the vector changes direction (i.e. the particle accelerates) from point to point. The central figure shows the particle at rest in the frame.

to the laws that describe its motion. The most famous equation tells us nothing about which form this rest energy takes in particular cases, but only that it is identical to the mass whatever form it has. More about what this means emerges from analysing the mass and energy of *systems* of particles. That rests on the foundation principles of addition and subtraction of spacetime vectors and is wholly geometrical. It is examined following the next paragraph.

4.4 Speeds, angles, rapidities

Given a constant finite limiting speed, then speeds in the same direction cannot be added arithmetically else they would inevitably add to more than c. SR has an addition theorem for velocities. It has the following form (since $c^2 = 1$ in natural units, it is suppressed in the last member of this equation):

$$u \oplus v = (u+v)/(1+(uv/c^2)) = (u+v)/(1+(uv))$$

That last member has the same form as the addition of angles has in hyperbolic trigonometry:

$$\tanh(u+v) = [\tanh u + \tanh v]/[1 + \tanh(u)\tanh(v)]$$

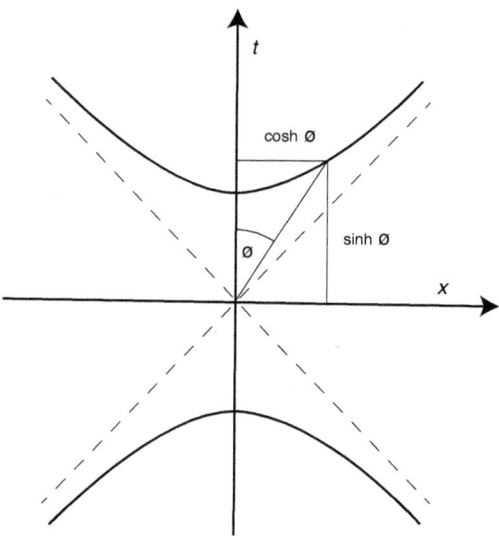

Figure 4.4: A moving particle's worldline is at the pseudo-angle θ to the t axis of the frame. It intersects the unit hyperbola yielding velocity components $\cosh\theta$ and $\sinh\theta$.

This reminds us that, in spacetime, velocity vectors differ only in orientation, i.e. by *angle*. Also a particle's motion relative to a frame is represented by a rotation through an angle relative to the frame's time axis. All the quantities and relations dealt with so far are best understood in terms of the trigonometry of spacetime angles. We can't add speeds but we can add angles and view the motion of particles relative to frames or to one another in terms of angle differences. Angles in this context are called *rapidities* to remind us that they are like speeds, but additive. Speeds are hyperbolic tangents of rapidities, dt is the sinh of the rapidity-angle and dx its cosh. Time, although indissolubly linked to space, is differentiated from it in the negative signs in the metric. So spacetime is called a pseudo space and the angles pseudo angles. But the hyperbolic trigonometry of spacetime is structurally just like Euclidean trigonometry and no more complex than it. This notation is not widely used so I will stick with $dx/d\tau$ etc. Finally on this topic, here is a further suggestive parallel: speed is like orientation in that the first derivative of speed is acceleration and the first derivative of orientation is angular velocity. Classically both combine with the second law of motion to require a force.

4.5 Mass and energy in systems of particles

The simple geometrical derivation of $e = mc^2$ (i.e. that rest energy is proper mass) explains in the simplest and most direct way why the equation is true in spacetime. For more about its meaning we must turn to the geometry of energy-momentum vectors for conservative interactions among systems of particles.

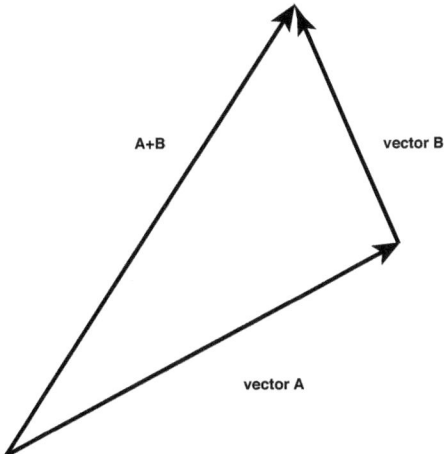

Figure 4.5: Vector $A + B$ is the vector sum of A and B. The components of the sum of two vectors is the sum of the components of the two vectors. In the Euclidean figure the magnitude of $A + B$ is always less than the sum of the magnitudes of A and B. In hyperbolic geometry it is always less.

Momentum and energy in classical physics are conserved in interactions. This is something like a conceptual truth, at least for energy, since that is why it was worth defining. Conservation principles also hold for the new energy and momentum in interactions both among objects and systems of objects. They rest on familiar symmetries and on the geometry of energy-momentum vector addition and subtraction. In the examples of interaction that follow I assume that the masses of system-constituents (objects at some chosen level of analysis) do not change i.e. where there are no unspecified sources or sinks of mass or energy.

It's important to grasp that vector magnitudes add vectorially, not arithmetically as quantities (masses) do. Vector addition can be described pictorially by the nose to tail operation, as shown in the figure. They can also be added by adding like coordinates However, unlike the Euclidean vectors shown in the figure, the spacetime hyperbolic geometry of 4-vectors means that, in addition, the magnitude of the resultant vector is greater than the arithmetic sum of the magnitudes of the summand vectors. The difference in signs in the signature, $(+---)$ of the spacetime metric (i.e. $ds^2 = +c^2dt^2 - dx^2 - dy^2 dz^2$) underlies the change.

4.6 Conservation in collisions

Take a familiar scenario for illustrating conservation of momentum and energy. It is expressed in two styles:

> *In frame F*, two equal-mass, (m) objects approach each other along the x axis, with constant velocities **v** and $-\mathbf{v}$. The total momentum of the system is 0.
>
> *In spacetime:* Each has an energy-momentum vector, constant in magnitude at each point on its worldline. They differ in direction and add to a system vector that has only a time component in F.

Both energy and momentum of the *system* of objects are to be conserved in interactions. The sum of the object momenta is $m \times ds/d\tau + (m \times -ds/d\tau) = 0$; the kinetic energy sum is similarly 0. The object energy-momentum vectors are at a rotation pseudo-angle, θ, to each other, so the energy-momentum vector of the system of these two objects is the *vector* sum of the object energy-momentum vectors. Since object vectors are not parallel, the system vector's magnitude is *greater* (shown as *less* in Euclidean figures) than the arithmetic sum of object vector magnitudes. That means that the mass of the system is greater than the arithmetic sum of the masses of the constituent particles. The conserved system-vector encodes these object vectors before and after any interaction.

Let us first suppose the collision is *elastic* (figure above). The physics of elasticity is not relevant to conservation. In this example, symmetry (the Minkowski/Ignatowski symmetries) requires that the objects differ only in the signs of their velocities after the collision, given that this was so before. The energy-momentum vectors of the rebounding objects retain the same pseudo-angle to each other and

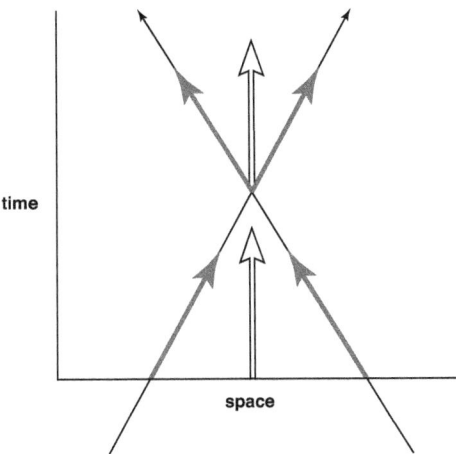

Figure 4.6: Elastic collision of two particles. Dark arrows are energy-momentum vectors for particles, the open arrows the energy-momentum vector for the system, the lighter lines the worldlines of the particles.

to the system energy-momentum vector despite their change in direction when they rebound. That is, the energy-momentum vectors of the particles after collision will differ only in a sign-reversal of the pseudo-angle each makes with the other: their sum, the system energy-momentum vector, is unchanged. Relative to F the system is, overall, at rest with zero momentum and kinetic energy; the kinetic energies of the rebounding objects remain what they were. Spacetime geometry entails conservation of system kinetic energy and momentum by the rules of vector addition: the components of the system vector will be unchanged in any frame.

Suppose a more interesting case where the collision is *inelastic*. The objects merge as a blob, stationary in the frame. The sum of the momenta, which now becomes the momentum of the stationary blob, remains 0. The sum of the *kinetic* energies now merged in the blob, becomes 0. Yet the *total* energy of the system is conserved since it always was the magnitude of the system-vector sum of the two energy-momentum vectors of the parts. The time component, now the only component, of the system energy-momentum vector remains unchanged: given that it is total energy, energy is conserved.

The magnitude of the *system* energy-momentum vector remains its mass, but now the system is just the blob. One object has the system mass. Both before and after the collision, the system mass is the vector sum of the object energy-momentum vectors, by symmetry. The system is differently configured in the elastic and inelastic collisions; nevertheless, whether configured either as the blob or as the rebound-

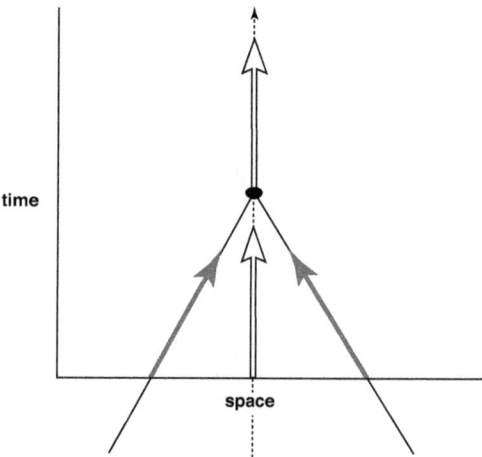

Figure 4.7: Objects and system energy-momentum vectors in an inelastic collision.

ing objects, *its mass is greater than the arithmetic sum of the masses of the objects before collision* (the arithmetic sum of the magnitudes of the object vectors).

The surprise in these examples is that the system mass is always greater than the arithmetic sum of the object masses. But that is simply a *geometrical* fact about the adding of vectors as against the adding of scalars. We need to keep tabs on what is added to what and how. Otherwise it looks as if something bafflingly material has happened – something, energy or mass, is lost, gained or destroyed. So I spell this out further.

Although the objects are simply stuck together in a non-elastic collision, we can choose to consider them separately. The mass of each is not what it was since their energy-momentum vectors are not what they were. Each vector is got by vector subtraction from the system energy-momentum vector. But now the vectors, both for the system and for each of its equal parts, are all *parallel* since the objects are stuck together, and are at rest in the frame. Parallelism makes vector subtraction the same as arithmetical subtraction. Thus each part has half the system mass and therefore is *more massive than before the collision*. The kinetic energy of each is now 0. The objects will be hot, naturally, and we may think of the heat as the motion of constituent particles gained from the collision. But the geometry does not say this: it simply tells its geometric story about vector addition and subtraction, not where energy has "gone" or "come from" or what form it has taken. The main story is not about matter physics, although, of course, there is some story to be told about the form of potential energy for the inelastic case.

Subtracting single object masses from system masses is equally surprising when the object energy-momentum vectors are not parallel. Suppose an object springs apart into two identical bits that fly off in opposite directions at the same speed. With respect to the original object's rest frame, the parts have equal and opposite momenta and equal kinetic energy after they spring apart. What is the mass of each bit? To find this, find the velocity 4-vector of each and find the magnitude so that the parts add vectorially to the magnitude of the system vector. Each has the same mass as the other but the arithmetic sum of these is less than the mass of the original object. Mass is "lost" and kinetic energy gained for each.

All this is a matter of how we *regard* lumps of stuff – as systems or as independent parts of them. When a nuclear device explodes, it expels radiation, particles and so on. The system mass and energy remain unchanged so system mass is not lost, but the story looks very different when told in terms of the outward-bound constituents. Mass seems to have vanished in violent puff of smoke, and energy is magically released instead. But it is the arrangement of parts that is so much changed. To understand how masses and energies of the outward-bound bits are determined is simply a matter of regarding them now as no longer a system but as distinct parts.

Something changes – an explosion happens – but no energy is gained or lost. The arithmetic sum of all the masses will be less, but that is about how to add and subtract vectors, not about anything vanishing out of existence.

Nature does not choose what is a system and what is not. We do. We can treat a warm body on the lab bench simply as warm but without kinetic energy, weigh it for its mass and so on. Or we can consider it as a beehive of swarming things each with its own mass and kinetic energies parcelled out among the bees. Mass is inertia, not quantity of matter. The energy-momentum vector of the system remains a vector sum: we have simply employed a microscope and exploited the structure of spacetime to get an accurate picture of the thing in bits. We havent changed it. We are studying geometry, not matter physics.

This completes the main case for the claim that $e = mc^2$ is principally a geometrical consequence of the natural representation of SR mechanics in spacetime. So are the phenomena of energy-mass "conversion." The simplicity and power of the spacetime representation of these phenomena makes a strong case in SR for the prominence, the power and the reality of spacetime.

4.7 Radiation

Massless radiation (light, photons) needs comment. The photon has energy i.e. frequency: but it has no mass. Since mass is the timelike magnitude of the *rest* energy momentum vector, a massless particle's worldline must lie in the null cone. Null vectors have 0 magnitude i.e. 0 mass. Light occupies the null (light) cone because it has zero mass rather than because it is the fastest signal. Relative to any frame, a photon moves a finite distance in a finite time [equal in natural units]: so a photon energy-momentum vector has *components* both of momentum and kinetic energy just as timelike (non-0 magnitude) energy-momentum vectors have. Since light has the same speed in every frame the magnitudes of its energy and momentum components are also the same. There is no frame in which they are 0.

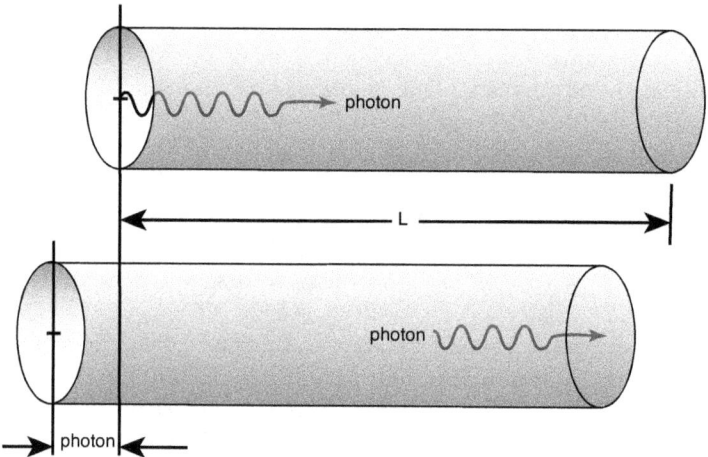

Figure 4.8: A photon is emitted from one end of a hollow cylinder and later absorbed at the other end. It recoils after emission and returns in recoil after absorption.

Radiation looks paradoxical. In a thing's rest frame, consider the emission of a photon. Before it radiates, the thing has rest energy and mass. It does not include photons as constituent parts, since 0 mass particles cannot be at rest. When it emits a photon, what leaves it has no mass but it does have energy and momentum. The thing recoils as the photon leaves. There is now a system comprising the thing plus the emitted photon. That is geometrically inevitable, given that there are photon 4-vectors. Since energy is conserved, the energy-momentum vector is unchanged but is now the vector of a *system*. It is parallel neither to the photon 4-vector nor to the thing's energy-momentum vector. The energy-momentum vector of the thing is given by subtracting the photon energy-momentum vector from the system energy-momentum vector. Its magnitude is less despite the magnitude of the photon-vector being 0. So although the radiation subtracts only energy from the system, the mass of the thing is less than that of the system. (In the next figure, the photon vector lies in the null (light) cone).

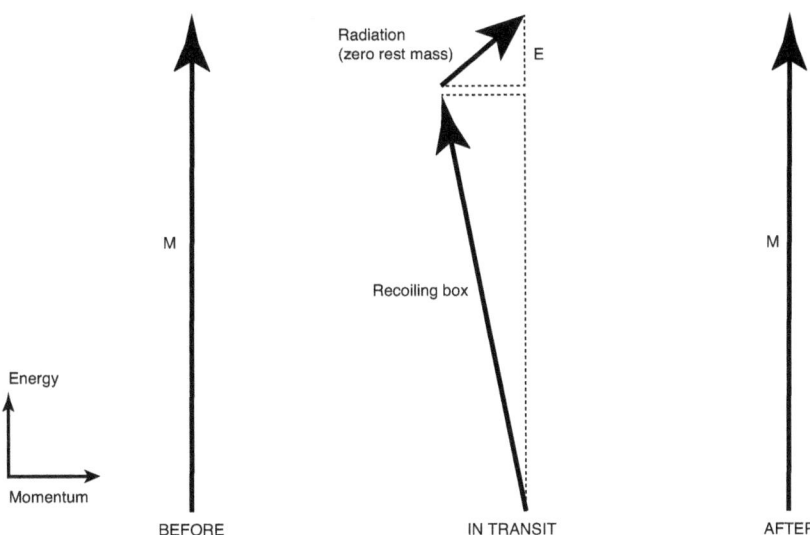

Figure 4.9: Energy-momentum vectors for the example of a cylinder and photon

Geometry does not entail that there are particles either of 0 or non-zero mass. It does not entail that the energy of light is its colour, its frequency. They are distinct, brute, facts, analogous to the facts that the value of c is what it is and that it happens to be the speed of something – light. It does not spring from the structure of matter theories. More cautiously, our present understanding does not allow us to say that it does.

5 IN SR, WHY DOES A MOVING ROD CONTRACT?

Chapter 5 examines the contraction of a moving rod in SR. It is both an exposition of this feature of SR and also a defence of one view in a recent controversy about the role of spacetime in explaining it. The noted physicist J. S. Bell wrote a paper arguing that it is explained by the way rods are made up rather than being a consequence of the structure of spacetime. This argument is examined in detail and some confusions and oversights in it are exposed. The whole critique also reveals how Minkowski spacetime gives a deep constructive explanation of this phenomenon.

5.1 Introduction

In the last decade a new explanation of the contraction of a moving rod, and of the slowing of a moving clock in Special Relativity has emerged in the work of Harvey Brown and Oliver Pooley (Brown, 2005), Brown and Pooley (2001), (2006). They describe it as "the dynamical interpretation of Special Relativity" (2006:77). It has created great interest and attracted both support and opposition.[1]

Their ambitions for this theory can be roughly made out first by citing the title of their joint paper of 2006: "Minkowski space-time: a glorious non-entity"; then from the following passage in Brown (2005:8):

> ... a moving rod contracts and a moving clock dilates *because of how it is made up and not because of the nature of its spatio-temporal environment*. Bell was surely right.

---

[1] For arguments opposing the dynamical interpretation but with different approaches from mine see Balashov and Janssen (2003) and Janssen (1995) (2009), Lange (2013), Norton (2008).

(Brown's italics)

The dynamical interpretation draws significantly for its explanation of contraction on two older texts, one by Einstein (1919) and one by a stellar figure in 20th century quantum mechanics, J. S. Bell (Bell, 1987). Bell's high status alone merits an interest in his views. His paper "How to teach Special Relativity" is the only contribution in Bell (1987) to relativity physics. This chapter examines Bell's argument both as a self-contained essay and as a basis for the explanatory style of the dynamical interpretation. I will argue that Bell misunderstood relativity and that he has been misunderstood, in turn, by Brown and Pooley.

5.2 On constructive explanation[2]

In the following passage, Einstein expressed a reservation about the 1905 Special Theory: it does not give a *constructive* theoretical explanation of its own peculiar "group of natural processes," the contraction of a moving rod being prominent among them.

> We can distinguish various kinds of theories in physics. Most of them are constructive. They attempt to build up a picture of the more complex phenomena out of the materials of a relatively simple formal scheme from which they start out. Thus the kinetic theory of gases seeks to reduce mechanical, thermal, and diffusional processes to movements of molecules – i.e., to build them up out of the hypothesis of molecular motion. When we say that we have succeeded in understanding a group of natural processes, we invariably mean that a constructive theory has been found which covers the processes in question.

> The advantages of the constructive theory are completeness, adaptability, and clearness, those of the principle theory are logical perfection and security of the foundations.

> The theory of relativity belongs to the latter class (Einstein, 1919:1).

I take the example as clearer than the definition. A constructive theory is a reductive one: the things and concepts of a surface or phenomenal theory are replaced in the explanatory theory by different, simpler ones. At the surface, things and their properties do not feature in the underlying explanatory theory. The ontology and ideology

---

[2]I am indebted to discussions with Syman Stevens in this section.

of the explanans are new. So the kinetic theory of gases is a constructive explanation of phenomenal thermodynamics. The kinetics of the molecular constituents of a box of gas explain what its heat, temperature and pressure are in terms of the motions of molecules, none of which has any phenomenal property. Further it explains the second law of thermodynamics as a statistical consequence of these motions given that the number of molecules in the gas is very large. The law may well fail for brief intervals of time and for closed systems with few molecules.

Other familiar examples are solidity, sound and colour. In the Introduction to his 1927 Gifford Lecture, (Eddington, 1928) Eddington wrote of his two tables, the table of ordinary phenomenal experience and the table of scientific theory. The former is solid, the latter made up of a swarm of rapid particles, none of them solid. The "occupied" space is largely empty. No particles touch or collide as phenomenal solids seem to do. Nor are the particles coloured. There must be many of them, but probabilities do not figure so strongly in these explanations as they do in the kinetic theory of gases.

If we are led by these examples, then the core criterion of a constructive explanation is that it is reductive: phenomenal things and properties are displaced at a new level by more basic theoretic ones. That the new entities and properties may, but need not, be micro physical is a merely accompanying feature and the reducing theory may, but need not, be statistical.

Constructive explanation is a common road to understanding some "group of natural processes" but it is not claimed to be the only one. Constructive explanations are not causal. My crumpled fender is caused by a bump from another car. There's nothing reductive in that explanation. By contrast, the total kinetic energy of a gas does not *cause* its heat content to have a certain value: its total kinetic energy is its heat, but *reconceived*. Less strictly, we might say that the motion of molecules causes the outcomes that, coarsely perceived, we call heat, pressure and so on.

Constructive explanations, then, are not principally bottom up as against top down. Aphoristically the main criterion demands that the phenomenal things and properties don't go "all the way down." At some level of analysis, new entities and new properties do the work.

What is the importance of the microphysical? It is common in constructive theories, but it does not displace the need for a reduction. To explain why blood is red we might say that it is not red all the way down. At a microscopic level we find platelets, white and red cells afloat in clear plasma. So blood is not really red although the colour of red cells predominates. That helps us understand the make up of blood at the bottom level but it tells us little about its redness. It is not a reductive explanation although it is microphysical in the sense of revealing hidden parts. It does more than simply multiply the number of constituents, since the parts are revealed as dissimilar. If descent to the microscopic level showed only that plasma and all its contents were red, that would be microphysical and bottom up but nevertheless a weak or perhaps no explanation: it subsumes the phenomenon under a covering law – your drop of blood is red because all drops of blood are red.

Many examples of constructive explanation have some statistical feature. The solidity of the desk needs a multitude of non-solid particles. This is not criterial.

Einstein did not point toward any constructive explanation for the contraction of a moving rod, nor indicate what one might be like, nor directly urge that we should find one. It would be a theory in which concepts of length, motion, rod, contraction and clock-rate are replaced by other entities and properties that explain how and why the contraction of rods in motion is phenomenal. There is no such *microphysical* theory consistent with Special Relativity. It is hard to imagine one that would be. Nevertheless, Einstein clearly hankered after it.[3] He was concerned about the dependence of the 1905 theory on rods and clocks, since these are alien to the main ideology of the theory. Later work on the observational basis has featured the role of light rays and free particles instead (Ehlers et al., 1972).

5.3 Bell's Lorentzian Pedagogy

I now turn to Bell (1987) "How to teach Special Relativity." Since its aim is pedagogical it is about the theory's *content* rather than its truth. I take Brown and Pooley to cast it as a constructive theory within SR for the explanation of contraction, but their own discussion is not confined to issues of content. Bell arrives at his exposition by invoking a Lorentzian pedagogy to reveal more clearly the reality underlying contraction in SR. Bell was an outstanding physicist. If he was wrong about this issue, as I shall argue, it is likely that many others share his confusions. First, we need to see what this pedagogy is and how its explanation is intended to work.

Bell saw a problem with Special Relativity: for instance, the relativistic contraction of a moving rod is often taught so as to lead students into error. But relativity's orthodoxy, as formulated by Minkowski

---

[3]Balashov and Janssen (2003:§4, 331-2) recognise the structure of Minkowski spacetime as providing a constructive explanation of spacetime, as I do. However, they do not mention reduction as the key aspect of constructive explanation although Einstein's example implies that it is. Their view is that constructive explanation rests on having a *model* of the theory. Without a further account of what a model is and how the having one explains phenomena, this sheds little light on matters. Models are not, in general, explanations.

in 1908, and thus widely understood today, sees no problem. Orthodox readers, me among them, don't easily grasp what Bell was driving at. He deflects attention outside relativity into pre-relativistic understandings of contraction. Yet his Lorentzian pedagogy is specifically aimed at teaching SR. If orthodoxy is correct and there is no problem, the Lorentzian pedagogy lacks point. That makes its interpretation difficult.

As an exposition of relativity's orthodoxy this chapter has say nothing new to say. As a speculative exposition of Bell's uncertain text it must labour over what seemed obvious to its author. That is unfortunate, but important issues demand it. Bell's argument is mostly about the contraction of moving rods. He also touches on the slowing of moving clocks but he raises no significantly different issues there. In any case, these particular features are only aspects of what needs to be explained and understood, for the deeper and more general issue is how to understand and explain the whole range of symmetries and reciprocities, of invariants and covariants, that emerge from relativity's transformations among members of the privileged set of inertial frames. But a restriction to just one concrete feature, such as "the contraction of a moving rod" will focus discussion and lose no significant generality.

I will describe two views of these contractions then sketch Bell's paper.

The *Fitzgerald contraction* as mooted by Fitzgerald and Lorentz, is a pre-relativistic concept: a rod, when in motion, is contracted as a function of its speed and in the direction of its velocity. Thus it is a state intrinsic to the rod (Fitzgerald, 1889); (Lorentz, 1892). This needs clarification as to what it is for a rod to move, to have a speed and to have a velocity. If the ether is rejected, is it motion through absolute space in the notoriously problematic style of Newton? A decision is needed here.

Further, a change in an intrinsic property of a rod requires a cause if it is not to breach the highly intuitive thesis of determinism. Motion through the medium of the ether took a step towards providing a cause although it has remained unclear how to work this through. Bell makes no appeal to motion through the ether nor does he plainly identify a cause.

In contrast, orthodoxy sees the *Lorentz contraction* as a function of motion relative to some inertial frame. A rod, uniformly moving with respect to a frame, is contracted in the direction of its motion. The contraction is relative to the chosen frame: it is a function of the relative speed. It is not intrinsic to the rod, and is a purely kinematical phenomenon.

By "purely kinematical motion" I mean motion that is force free. But kinematics is strictly the study of motion *abstaining from a study of forces* that may cause it. The distinction matters somewhat in §5.12.

Throughout, Bell uses the phrase "Fitzgerald contraction;"[4] "Lorentz contraction" never occurs. Why? Fitzgerald's discussion of the contraction came before relativity, so it has a priority. Perhaps Bell was reminding us of that. But Fitzgerald's view of it differs sharply from that in orthodox relativity. Was Bell reminding us of that – and recommending a return to it? I shall argue that he was and did.

5.4 What Bell said

I distinguish four parts of the paper.

(i) The first page opens strongly. Bell's teachings "emphasize the continuity with earlier ideas" and "play down the radical break" of special relativity. The radical break may "destroy completely the confidence of the student in perfectly sound and useful concepts already acquired" (67). Admiring references to the work of Lorentz, Fitzgerald and Larmor pervade the paper.

(ii) Bell supports these claims by discussing a remarkable misunderstanding of the contraction of a moving rod. He describes confusion in the canteen at CERN over the "old" puzzle of a taut thread between two rockets in a nose-to-tail alignment. Will the thread break if the rockets are accelerated identically and simultaneously relative to the rest frame? I discuss the example in §5.8 and §5.13 below.

(iii) There follows a deduction of the contraction from Maxwell's equations, following the approach of Lorentz and Fitzgerald. It aspires to explain the contraction "without mystification." This is the centrepiece of the paper. I quote from Bell's opening outline (69):

> For a charge at rest ... the familiar Coulomb field [is] symmetrical about the source. But when the source moves very quickly ... the field is no longer spherically symmetric ... roughly speaking the system of lines of electric field is flattened in the direction of motion ... therefore a body set in rapid motion will change

---

[4]The verb phrase "Lorentz contract" occurs just once in his paper (75). "...in the rocket problem of the introduction, the material of the rockets, and of the thread, will Lorentz contract. A sufficiently strong thread ... would impose Fitzgerald contraction [sic] on the combined system." Grammar suggests a reading in which the phrases refer to the same contraction; or, implausibly, that the one contraction is imposed on the other.

shape. Such a change of shape, the Fitzgerald contraction was in fact postulated on empirical grounds by G. F. Fitzgerald in 1889.

(iv) The last page is conciliatory. Lorentz and Einstein differ, in "two major ways. There is a difference of philosophy and a difference of style ... the facts of physics do not oblige us to accept the one philosophy rather than the other" Philosophically, "Einstein *declares* [my italics] the notions "really resting" and "really moving" meaningless ... only the relative motion ... is real." Lorentz "... *preferred* [my italics] the view that there is indeed a state of *real* [Bell's italics] rest defined by the "aether"..." (77). Einstein sounds arbitrary and peremptory, Lorentz arbitrary and whimsical.

He goes on "... the special merit [of the Lorentzian pedagogy] is to drive home the lesson that the laws of physics in any *one* reference frame account for all physical phenomena, including the observations of moving observers. ... in my opinion there is also something to be said for taking students along the road made by Fitzgerald, Larmor, Lorentz and Poincaré. The longer road sometimes gives more familiarity with the country."

5.5 A conjectured interpretation

Bell's *only* explicit complaint is that, in teaching SR, the confidence of students "in perfectly sound and useful ideas already acquired" is often completely destroyed. He neither specifies what these already acquired ideas are, nor just what the radical break is, nor how the break does its damage. He claims (68) to deduce the contraction from Maxwell's equations "without mystification" following the approach of Lorentz, Fitzgerald, and Poincaré; he does not describe a mystification, nor how he avoids it.

Of these remarks, one – "destroy completely the confidence of the student in perfectly sound and useful concepts *already acquired*" (67, my emphasis) – seizes attention. The italicised phrase, rather straightforwardly, refers us to pre-relativistic concepts. Which could these be? Among earlier conceptions of the contraction of a moving rod were those of Fitzgerald and Lorentz as described above. The context, the words quoted and Bell's insistence on the phrase "Fitzgerald contraction" strongly imply that he had just that in mind. It strains credulity that he was so grossly careless in framing his one explicit objection to orthodox teaching that he confused the Lorentz with the Fitzgerald contraction. However, Einstein's (1905) and modern orthodox relativity *does* completely undermine confidence in pre-relativistic concepts! Exactly that was its great achievement. Even if Bell did not have Fitzgerald in mind, earlier relevant ideas were, without exception, those of a contraction that is intrinsic and caused. If so, "A moving rod contracts" is semantically like "A cooling rod contracts:" i.e. no reference to another relatum is relevant. Lorentz himself mentions the analogy with heat expansion. That is not a concept within Special *Relativity*!

When discussing the accelerated thread example, Bell writes:

> ... as the rockets speed up [the thread] will become too short because of its need to Fitzgerald contract, and must finally break. It must break when ... the artificial prevention of the natural contraction imposes intolerable stress. (67) ... A sufficiently strong thread would pull the rockets together and impose Fitzgerald [sic] contraction on the combined system. (75)

That language fits ill with the orthodox account; it is naturally read as implying some kind of intrinsic and therefore absolute contraction. Brown's *"because of how it is made up and not because of the nature of its spatio-temporal environment"* also suggests an intrinsic contraction, not a relativistic one. Neither writer is quite explicit. Nothing in their texts forbids intrinsic contraction. Surprising as this interpretation may be, it is the most natural reading of Bell's text. I doubt that it is a straightforward interpretation of Brown.

However, if determinism is to be retained, an absolute, intrinsic contraction needs a cause. It needs to be dynamical. Without a cause, the contraction is mystified. I argue that Bell did go on to offer a causal, dynamical explanation and probably saw it as dissolving his unspecified "mystification." Such an account might be *true*, but it would not be *relativistic*. It would give a causal, not a constructive, explanation.

On the last page, there is nothing about radical breaks or how confidence is restored in sound concepts already acquired. One is thus reminded that, even on the first page, Bell writes only of "emphasising the continuity with older ideas" and "playing down," rather than correcting, a "radical break." So it seems that he did take himself to be explicating SR.

How could he think so? In the end, one must make a best guess among alternatives each of which is unlikely on the face of it. I take the following story as optimal.

The Lorentz transformations entail, relative to an inertial frame, contraction of any rod in uniform force-free motion. In that purely kinematic setting, contraction is mystified, one might think, by explaining it, in the rest frame, through fixing the simultaneous locations of the end points of the moving rod. That is not a direct measurement of the rod. Further there is no mention of a cause. This is unsatisfactory if the contraction is intrinsic to the rod. I assume that Bell meant by the "radical break" these features of the Lorentz transformations.

There is a more radical candidate for the break: the departure from motion-through-the-ether theory as both cause and explanation. Bell's emphases on Fitzgerald and Lorentz strongly suggest it.[5] However, on the last page, Bell dismisses this as philosophical; physics does not demand that we choose between Einstein and Lorentz. That puts the more radical candidate out of consideration. Further, teaching an ether theory is plainly not teaching SR.

The last page references to the "special merit" and the Lorentzian "longer road" refer us to the centrepiece deduction from Maxwell's equations, a deduction "without mystification." It also deduces the observations of moving observers from these equations. Thus it plays down the Lorentz transformations by making them derivative and secondary.

5.6 The question of acceleration

How did Bell think the Lorentzian pedagogy teaches SR despite its contraction's being intrinsic? Plausibly, he saw acceleration – not the motion but the *setting* in motion – as the key. Both here and in the breaking-thread problem in part (ii), acceleration provides the *context*. (See the quotes in §3.1 above.) Unless he saw acceleration as having this role there was no reason for mentioning it at all.

Two aspects of acceleration are relevant. First, non-uniform motion is absolute: a body in non-uniform motion relative to one frame is in non-uniform motion relative to all. Contraction as arising in acceleration looks like yielding an absolute, real, intrinsic change. But caution is needed: acceleration, unlike non-uniform motion, is neither Lorentz invariant nor absolute. Second, acceleration is dynamical, caused. Thus the thought seems to be this: the cause of its acceleration – what sets the rod in motion – acts on the structure of the rod so that this contraction is intrinsic and absolute. Bell does not say these things, but both are, at the very least, consistent with his discussion and something like them is rather clearly called for by it. Thus Brown (2005:124) (2005) takes Bell's question to be: "what is the prediction

---

[5] I am indebted to Michel Janssen for pointing to a possible connection between Bell's non-committal stance on this aspect of his Lorentzian pedagogy and his proposal that a return to Fitzgerald, Lorentz, and Larmor might be the price we have to pay for instantaneous measurement collapses in QM. Perhaps that was part of Bell's agenda. See Janssen (1995:§2.3.5, note 63)) and Balashov and Janssen (2003:336). Bell made this proposal in a well-known interview published in Davies and Brown (1986:49).

in Maxwell's dynamics ... as to the *effect* on the electron orbit when the nucleus is (gently) accelerated in the plane of the orbit?" (my italics). If this does not directly imply, it does strongly suggest, that the gentle cause of the acceleration also causes the effect on the electron orbit.

So on this interpretive hypothesis, Bell wished to show that, when we turn to acceleration, to real absolute changes and to their causes, we see contraction in a different light from the hasty, purely kinematical, light in which the orthodox relativist sees it. Here are real facts of physics: we see, in the caused, bottom-up changes in the Coulomb field, through the electron orbit to the whole rod's contraction that the process really is intrinsic and really caused: we see something like a causal chain. The "longer road" has the signal merit of showing how and why it develops. Since we need no preferred frame to see all this, it *is* SR in a new perspective, freed from mystifying restrictions to inertial motion.

There is a problem with this. Clearly motion alone, if it is merely relative, cannot cause any intrinsic change in the uniformly moving thing. Nor can *non*-uniform motion bring about any well-defined intrinsic change, since acceleration is not Lorentz invariant. But since acceleration *has* a cause, its cause is at least a candidate for causing contraction in an accelerated rod, thus rescuing determinism. However, all that Bell shows about the micro-contractions is that they follow from setting the rod in motion. No cause is specified so the micro-changes are as mystified as is the contraction of the rod itself. We need a cause. The only one in the offing is the accelerating cause. Plausibly – more plausibly than any other reading – Bell saw it as the cause of the contraction.

5.7 What explanation lies at the end of the "long road"?

At the end of the long road (74) Bell sums up what he claims to have shown, subject to two provisos: "Can we conclude then that an arbitrary system, set in motion, will show precisely the Fitzgerald and Larmor effects? Not quite." Provisos are not main conclusions, but merely modify them. The main thrust of his explanation is *already given* just in the deduction from which he draws the conclusion he mentions. It is that the acceleration, or the accelerating forces, cause the contraction. The provisos, being provisos, merely, tidy up.

The first reservation about the explanation is that "the Maxwell-Lorentz theory provides a very inaccurate model of actual matter." Bell lists several problems; the proviso follows them (74): "We need not get involved in these details if we assume with Lorentz that the *complete theory* is Lorentz invariant ..." (original italics). That assumption plays no role in the explanation. The dynamical interpretation of Brown and Pooley, citing just the Lorentz covariance of all matter theories, certainly figures in Bell's paper, but, at the very least on the face of it, *not* as his *explanation of contraction*. It is an assumption that allows us to skip dynamical complexities as distractions and turn to the explanatory deduction itself. In that sense, he was not concerned with "how the rod is made up" but with how contraction is caused. His explanation is not constructive in any sense, let alone in a reductive one. Nor did he say it was.

Bell also tells us that, to get just the Fitzgerald contraction, we must avoid cases where the acceleration is so violent as to destroy the rod. So he limits his explanation to cases of gentle acceleration, i.e. gentle force. But the limitation is irrelevant.

Arbitrary systems set in motion, gently or not, never show precisely the Fitzgerald contraction. There is a single exception: Rindler acceleration (discussed in §5.13 below). Surprisingly, Bell has already provided a counterexample to his main conclusion – the case of the identically accelerated rockets and thread. Identical accelerations may be as gentle as you like but they will not show a Fitzgerald contraction in the thread: they will break it!

More importantly, relativity does not identify non-uniform motion relative to a frame, with acceleration rather than *deceleration*. Neither is a Lorentz invariant. Any non-uniformly moving rod will be decelerated, slowed and expanded relative to some class of inertial frames. This plainly entails that the change in shape of an accelerating rod is a change relative to a frame, not something intrinsic to the rod. Grant that and it becomes unclear what role dynamics can play in the matter. The "special merit" of the single-frame account blocks insight into highly relevant differences as regards other frames.

In sum, given a rather stable body in an inertial state, it is Lorentz contracted relative to every frame in which it moves. But, a non-inertial body, supposing it rather stable under its motion-changing force(s), accelerates and Lorentz contracts relative to members of only one subset of inertial frames. It decelerates and Lorentz expands with respect to all members of the infinite subset that is a complement of the first. These can't both be intrinsic.

Bell does not quite ignore the infinitely many ways a rod may accelerate from rest to some state of uniform motion relative to its initial frame, but, again, these complexities are not well described, and are included as a mere proviso to the main explanation. The relevant paragraph (75) concludes: "Thus we can only assume the Fitzgerald contraction, etc., for a coherent dynamical system whose configuration is determined essentially by internal forces and only a little perturbed by gentle external forces accelerating the system as a whole."

In general, gentle external forces on a rod will affect its length so as to swamp any Fitzgerald contraction. Think of the difference between gently accelerating a steel ruler by pulling one end of it as against gently pushing from the other. The different locations and directions of the pull and the push cause elastic expansion in the first case and contraction in the second. The changes in length in the rod will be minute; yet they dominate the relativistic contraction.

In common examples, forces that accelerate a rod perturb it from an equilibrium state. Once the perturbations subside, either a perfectly elastic rod will be in an identical equilibrium state relative to its new rest frame; or, if an inelastic rod, it will have some deformed equilibrium state. In relativity, the criterion for its complete escape from accelerating force distortion will be that, relative to its new rest frame, the now freely moving rod shows this perfect elasticity. The forces accelerating the rod are irrelevant to the Lorentz contraction.

Further, suppose that the accelerating forces are very strong. They are destructive only when they have very different magnitudes and directions at neighbouring points. Forces may be arbitrary in magnitude provided that their differences are appropriately graded along a body's length. (See §5.13.)

5.8 The puzzle of the thread

Bell intended the thread example in his part (ii) to show that things may go wrong when SR is taught in the orthodox way and perhaps to hint at how. He gives the example in an amusing and surprising story.

> This old problem came up for discussion once in the CERN canteen. A distinguished experimental physicist refused to accept that the thread would break and regarded my assertion, that indeed it would, as a personal misinterpretation of special relativity. We decided to appeal to the CERN Theory Division for arbitration and made a (not very systematic) canvass of opinion in it. There emerged a clear consensus that the thread would **not** break! (68; Bell's bolds).

Bell showed that the problem has a perfectly trivial solution in the rest frame. Accelerations of distinct points that are identical with respect to a frame, and start simultaneously, maintain the interval between them *in that frame*. That's what "identical acceleration" means. A thread thus accelerated will not contract relative to the frame: it will break. Bell says "it will become too short, because of its need to Fitzgerald contract, and must finally break" (68). But it is more illuminating to consider other frames. The rocket firings are not simultaneous relative to any inertial frame in motion relative to the first (assuming standard configuration). In any frame in which the rockets *accelerate* the complex of rockets-and-thread from its uniform motion, the front rocket fires first and thus pulls and breaks the thread. In frames where the firings *decelerate* the uniformly moving complex, the rear rocket fires first, thus slowing its end of the thread and snapping it. Here we see a clear causal role for the rockets. After all, only they are new causes in play.

This tells us that the CERN canteen and its Theory Division went wrong. It does not explain why so many very smart people stuck to so bad an answer.

I guess they thought like this: in uniform motion, all parts of the body move together as a whole and partake of the same motion. They all have the same velocity. Next, assume that the general case of non-uniform motion is *"accelerating the system as a whole"* (75), a matter of all parts of the body accelerating somehow together. Finally think that all parts being accelerated together means their being identically accelerated. That prompts the quick-draw solution that identical and simultaneous acceleration entails just the Fitzgerald-Lorentz contraction and the intact thread. Bell did not take the last step but it is clear (75) in the phrase I italicised above, that he took the earlier ones. He thought that we can usefully speak of accelerating the system as a whole, so long as it is gently done. The question how this would differ from identical acceleration is left untouched despite its obvious relevance.

Non-uniform motions are more complex. Even in the case of overall inertial motion, things need not move so that all parts have the same uniform velocity. Suppose you are seated on a railway platform. A fast train passes at uniform speed; on board, someone plays a concertina. Its parts are all moving differently, some accelerating in different ways from others, some parts in different ways at different times. Nevertheless, there is a sense in which, as a whole, the train and its contents all partake of the uniform motion of the train. Each part of it, all the time, has a component, **v**, the relative velocity of the train, as part of its total composite relative velocity.

There is no counterpart of this in non-uniform motion. There need not be, generally won't be, the same non-uniform, constant component in the composite velocity of each part of a rod. The accelerations by external forces of each part must be separately given for each point of the rod and each moment of acceleration before we can picture "its acceleration." In general, acceleration (deceleration) vectors differ in magnitude and direction for each point.

Thus a close look at non-uniform motion shows that contraction in relativity theory is not among the perfectly sound and useful ideas already acquired before relativity's advent.

5.9 Difference of style

Plausibly, Bell saw his comments on the difference of style between Einstein and Lorentz as motivating the special merit of the one frame and the longer road: "The difference of style is that instead of inferring the experience of moving observers from the known and conjectured laws of physics, Einstein starts from the hypothesis that the laws will look the same to all observers in uniform motion" (77). This claim, repeated in Brown and Pooley (2006:262) is significantly false. The hypothesis is Einstein's first postulate, but it is not how he started. In 1905 he began with a problem. Contrary to Bell, it's a problem encountered in inferring the experience of moving (and rest) observers from the known and conjectured laws of physics. The first sentence of Einstein (1905) is: "Maxwell's electrodynamics as usually understood at the present time when applied to moving bodies leads to asymmetries which do not appear to be inherent in the phenomena" (Lorentz, 1923:37). Different conclusions about the same objects are drawn depending on their motions; e.g. an encounter between a magnet and a conducting coil is different if the magnet moves inside the coil, than if the coil moves to surround the magnet. This is easily rephrased in terms of different conclusions drawn by differently moving observers. In 1905, Einstein did not present a long road to this problem, but there is no reason to think he never travelled it. He must have done, to use it as he did. So the paper begins, not with postulating the invariance of laws, but with paradoxically different entailments for still or moving observers.

Another example sharpens Einstein's point. Consider a magnet at rest in a conducting coil. For an observer in that frame, there is nothing to produce a current in the wire. But there is for an observer for whom both are in motion. In that case, the motion of the magnet produces a current-inducing **E** field in the coil. But the motion of free electrons in the coil also produces an equal and opposite **E** field. So at each point there are equal, opposite, and therefore cancelling, **E** vectors. Thus neither observer measures a current in the coil, but the field values calculated by each are quite different. (See (Norton, 2010) for a clear, easy account and a diagram.)

No coil is both devoid of current-producing **E** fields and yet contains two such equal and opposing **E** vector fields. On the face of it, the coil is just a familiar physical object. But these conflicting field descriptions cannot both be true of one and the same familiar object.

Either, at least one of the observer's claims is false but nothing can tell us which one (as in an ether or an absolute motion theory); or the description is relativistic, and so not directly about a familiar physical thing but about a thing-in-a-frame. Phrases such as "the rod," "the coil" are referentially indeterminate in relativity theory. In the 1905 theory we don't quite know what we are talking about.

5.10 What is a relativistic rod?

Operations to measure the length-of-a-rod-moving-at-**v**-in-F_m leave the semantics of relativistic language somewhat up for grabs. There are several ways to parse, or de-hyphenate, its expressions. What is measured is the spatial interval between two events simultaneous in a frame. In another frame, the two events are not in any of its simultaneity-determined spatial intervals. Let "Excalibur" in ordinary speech rigidly designate my one, very own, rod: then Excalibur-in-F_m and Excalibur-in-F_n, are not identical referents despite the uniqueness of Excalibur: they have incompatible length properties. The *operations* that explain why these phrases are referentially indeterminate don't define or make explicit what *entity* is measured. Rods and coils are not basic entities. What things are? Seeking a referent for "rod in F_m" makes possible a route to a constructive explanation through a change in ontology and ideology. (See also Petkov (2009:Chapter 4.5).)

Thus there is a basic semantic gap in Einstein (1905) (before Minkowski's conception of spacetime). Einstein's first paragraph presupposes that some definite entities are basic in relativity, but does not specify them.[6] Neither Bell nor the dynamical interpretation mentions this highly relevant peculiarity.

Consistency in syntax is easy: Lorentz covariant properties must not be ascribed to a rod but only to a rod-in-a-frame; if a frame is contextually presupposed, "the rod" may refer uniquely. That roughly describes common practice. But the underlying semantics is less straightforward. That is given in the next sub-section.

Bell's Lorentzian pedagogy, confining itself to one frame, ignores the semantic issue that makes the reference of "the rod" indeterminate. The pedagogy leaves students naïve as to what the explanation is about – familiar rods or relativistic rods-in-frames. This is closely related to what we saw before that, in an appropriate frame of reference, a non-uniformly moving object may be slowed and will expand relative to that frame.

It is high time to summarise this examination of the Lorentzian pedagogy. The pedagogy aims to explicate the content of SR but

---

[6]Of course, this is not intended to describe Minkowski's route to spacetime.

it is about a different theory in which contraction is intrinsic and needs a cause. It is presented as a causal (dynamical) explanation of contraction, not a constructive one. It treats contraction as intrinsic to the rod, caused by the changes in its field values, these changes being initiated, presumably, by the cause of the rod's "acceleration as a whole." If my objections are sound, then we have no satisfactory account of how accelerating forces could do this so the pedagogy fails. It also fails to teach the content of SR.

5.11 What is SR?

There are at least four options how to pursue the best, the true, theory: (i) the path of relativity: this entails solving the referential problem of "rod-in-a-frame" (ii) add an ether to ontology, accept the contraction as intrinsic to ordinary rods and explain it as caused by its motion through (some variant of) the ether (iii) look deeper among the causes of acceleration for something that could cause the Fitzgerald contraction (iv) presupposing some version of Bell's account, bring contraction, time dilation and other covariances, invariances and symmetries under a broad covering law such as "dynamical theories are all Lorentz covariant." For (iii) and (iv) there remains the question about how motion can cause a rod's contraction[7] and how determinism is to be retained.

Each option has an ontological element, either in the entities it admits or rejects or where it finds or omits causes. Thus, in each, the world is physically different. Option (i) was taken by Minkowski (1908). (ii) also obliges us to accept a new, presently undescribed, object in our ontology. It needs well-defined properties to show how ether causes contraction. The task may not be impossible but the well-known epistemological costs of pre-relativistic ether theory are unwelcome. The world conceals both the ether and many intrinsic properties of things that move undetectably through it. Since we can define all classical physical properties (except charge) in terms of mass, length and time, the epistemic barrier between us and the world of intrinsic properties is virtually complete. Option (iv) appears to be a last resort in case other options fail. We leave through the same door by which we came in.

Minkowski (1908) provided a constructive explanation for the results of Einstein (1905) by moving to a new ontology and ideology. It explains not merely contraction in motion and other covariant phenomena but also explains and unifies all Lorentz invariances and covariances, together with the symmetries and reciprocities that make SR such a strikingly novel and comprehensive theory.

---

[7]The same question arises for all other covariances.

There are 4-dimensional objects, worldlines and worldsolids set in a spacetime of appropriate geometrical structure. This re-conceives the motion and contraction of moving rods as phenomena resting on the relative orientations of 4-dimensional entities and their 4-dimensional properties as these occur in Minkowski spacetime. That is then analysed by the frame-relative approach so natural to experimenters and theorisers who inevitably conceive themselves as agents and as 3-dimensional continuants. The metric of spacetime places familiar constraints and freedoms that give preferred status to Lorentz frames of reference. These new 4-dimensional objects do not move, they do not contract with motion therefore, and the slowing of clocks is a matter of the proper time metric along their worldlines. The new things and new properties yield a reductive explanation of the phenomenal properties of phenomenal rods and so on, as these appear in frames.

Dodging between spacetime and space-and-time languages, a referent of "rod" is a rather notional 3-dimensional continuant entity composed of a time-extended continuum of parallel spacelike sections of a 4 dimensional worldsolid – the 3-rod and the 4-rod, let us call them. A single spacelike section of a 4-rod is oriented within it, coinciding with its orientation in a frame: if time-orthogonally oriented, then the 3-rod is at rest in that frame. If not thus oriented then the 3-rod is in relative motion. So Excalibur-in-F_m really is not identical with Excalibur-in-F_n, since the 3-rods are composed of distinct spacelike sections of the same 4-rod. The two are distinct since their differently oriented spacelike-sections are distinct. But this springs also from differences of orientation among 4-rods absolutely and thus also within the same frame: the 3-rods have different velocities in any one frame. Thus full referential determinacy comes with phrases such as Excalibur-at-**v**-in-F_i. (Petkov (2009:Chapter 5.3).)

As for how 3-rods are "made up," we see, in different frames, differently oriented components of the same electromagnetic 4-tensor-field in play. So the **E** field in one frame has certain values, as mere x_m–components, of a 4-tensor. When the frame is changed from F_m to F_n (assuming a standard configuration) the new, x_n–components differ but without change in the condition in spacetime that the tensor field describes. That is the reality that underlies the phenomenal field changes deduced by Bell from Maxwell's equations. There is no frame-transcending 3-dimensional object (rod or coil) common to both frames. There is no frame-transcending dynamical (causal) description or explanation of the contraction of one and the same entity. The explanation of all this is purely geometric and purely kinematic. "Rod" is frame-indeterminate referentially: that is the price of consistency in relativity.

A main player in this reduction is spacetime, through its metric and affine structures, its orientability and the vectorial structures that its metric sustains.

It is not bottom-up explanation nor is it a model in the familiar sense. But it does explain the phenomena by appeal to an underlying, though not a microscopic, reality. The ontic change demands a new unity of space and time. This makes the theory's ontology more economical, since space and time, separately, preceded it. It is a geometrical explanation of the phenomena, not a dynamical one. The symmetries of transformations from one inertial frame to any other, reflect the structural symmetries of spacetime. (See Janssen (1995), (2009).)

5.12 The dynamical interpretation of contraction

I have argued that Bell understood his intrinsic contraction as being caused by accelerating forces, and that he was mistaken. I should emphasise that many of the preceding objections do not depend on that interpretation. Assume instead that Bell's longer road does not spell out a causal development culminating in a contracted rod. Then the flattening of the Coulomb field and the contraction of the electron orbit under acceleration, are left in the same boat as the contraction of the rod. They all have the very same contraction and if the one is mystifying, so is the other. If the dynamical interpretation does not draw on Bell's explanation as a causal process, and it does not tell a reductive story, it is not clear how it is a significant explanation. We can deduce the contraction from Maxwell's equations or from the Lorentz transformations. We can say that all our theories are Lorentz covariant. Any of these formulations tell us what happens; none gives a cause nor tells us why it happens. The dynamical interpretation can give no role to acceleration or its causes since Lorentz covariance concerns only transformation from one inertial frame to others. It yields no explanation at all in terms of an underlying reality that explains the Lorentz invariance of laws or the covariance of properties under Lorentz transformations or the importance of rest or uniform motion.

In the end, we have to guess what Bell's theory of contraction is. If my interpretation is correct, then the dynamical interpretation of Brown and Pooley does not appear to be a truncated version of it. It is not clear how they understand Bell's longer road or whether they regard contraction as intrinsic to a rod or relative to a frame. If the longer road does not provide a causal explanation then it merely gives a bottom up (aka an all the way down) account of the contraction. If the bottom up deduction is omitted in a truncated version of the pedagogy, then it is unclear what the dynamical interpretation's claim to being a constructive explanation has to do with Bell. The core message of the dynamical interpretation appears to be that, for SR, all fields are Lorentz covariant. Obviously the message is correct, but its source in Bell is merely the assumption he makes in the first proviso. Further, we learn that explanation stops at this message: it is to be accepted as brute fact (Brown and Pooley (2006:84). If we think it an implausible coincidence that all fields are Lorentz covariant, then we nevertheless concede that this is possible. But we don't thereby concede it any plausibility or admit that it is anything but an accidental "law."[8] The dynamical interpretation of Bell's Lorentzian pedagogy is not an illuminating explanation of why a moving rod contracts. Nor is it clear that covariances under spatiotemporal transformations that map from one inertial frame to others is, after all, not "because of the nature of its spatio-temporal environment" (Brown (2005:8). The explanatory power of Minkowski's reductive ontology and ideology is deeper and far more readily understandable. It is purely kinematical.

If it is somehow a matter of the dynamics of material structures, how would the dynamics determine it? The constitution of matter, including the length of a rod in an equilibrium state, is largely determined by electromagnetic forces. *That* is how a rod is made up. Maxwell's equations entail a wide range of observable phenomena and laws of behaviour. They explain electromagnetic induction, Gauss's law, the structure and nature of electromagnetic radiation and a great deal else. But how does this concern contraction as a dynamical change or show how the slowing of a uniformly moving clock is dynamical? Among the pervasive symbols in Maxwell's equations are the field variables, **E** and **B** and the constant c. **E** and **B** fields are dynamical: they represent the components of force vectors. But c represents nothing dynamical. It is a universal constant with dimensions metres/seconds: it is a kinematical element in Maxwell's equations. It merely happens to be the speed of electromagnetic radiation but only if the photon is massless. That is why light is the fastest, constant, invariant signal (if, indeed, it is). The significance of the finite maximum invariant speed, c, is not that it is light's speed. That was the argument of Chapter 3.

Maxwell's equations do not determine its value, yet length contraction and time dilation depend wholly on c. If c were infinite the Maxwell relations among **E** and **B** fields would be structurally unchanged: but there would be no contraction of moving rods or dilation of time. There would be no Lorentz covariances. Once c is finite, contraction is determined by the ratio of v to c. If c were some domestic speed – 10 mph in Gamow's early popularisation (Gamow, 1957) – the structure of electromagnetism would be just as it is, including the

---

[8]For an examination of the scope and model hierarchy of relevant laws see Lange (2002) and (2013).

structure of electromagnetic radiation and so on. But the contraction of a moving rod would become grossly observable. Changing c's value does nothing to change the dynamics of the theory but everything to change the contraction of a moving rod and of every other Lorentz covariant property.

Since we don't know why c has the value that it has, there is something we don't know about why moving rods contract to the extent that they do. In a similar way, constants in many physical theories are independent of the dynamic structure of their embedding. We don't know how to derive them. They are brute observational facts of physics in our present understanding. But if c lies in the open interval $0 < c < \infty$ we can explain, in kinematical relativity, not only why moving rods contract, but, quite generally, why mechanics and electromagnetism are Lorentz covariant.

5.13 Lorentz contraction and spacetime

Finally, and to illustrate and clarify what these arguments aim at, I consider the unique case where non-uniform motion yields exactly the Lorentz contraction or expansion, depending on which frame it is described in. I set the example in spacetime, where it can be made immediately obvious in a diagram. It also needs no dynamical variables to explain it. I follow usual practice in using "acceleration" in a spacetime context, to include frame-relative deceleration, thus referring to non-uniform motion absolutely. It will be explicit when the context is frame relative.

Let us remind ourselves that kinematics is the study of motion without studying its forces if any. It is not confined to the study of free motion.

Let us ask:

(i) Is there a pattern of non-uniform motion for a rod that, with respect to its original frame, results only in its Lorentz contraction (expansion)?

(ii) Does this pattern have the result that the rod has the same length with respect to every momentarily co-moving inertial frame?

(iii) Does the make up of the rod play any role in the outcome if it moves in this way? I.e. would a soft rubber rod behave differently from a steel or wooden one?

The first two questions are about only the kinematics (geometry) of the non-uniform motion.

Uniform acceleration, or hyperbolic non-uniform motion, of points in a rod answers the first two questions affirmatively. It was studied in

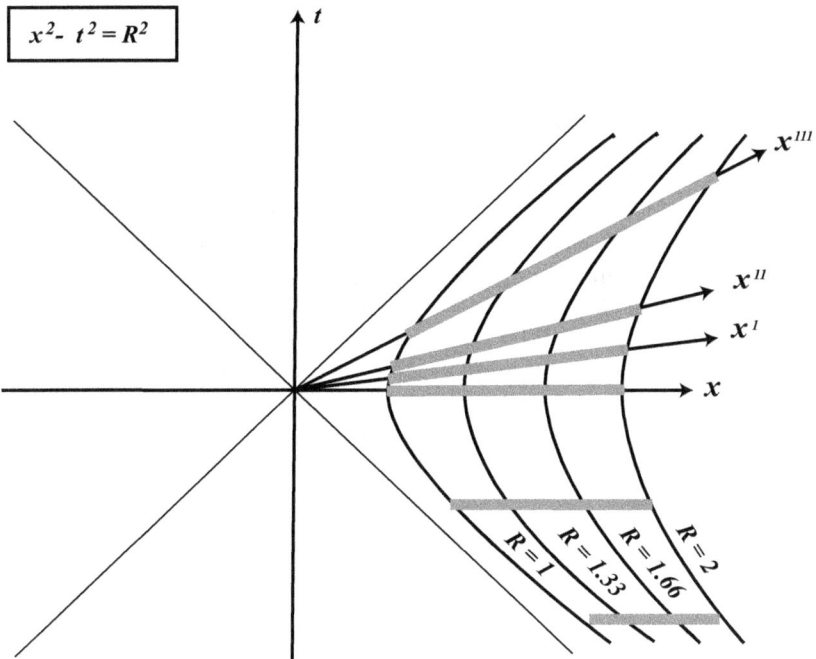

Figure 5.1: Uniform acceleration of points in a rod. In the lower right quadrant, are shown two positions for the rod relative to the frame of reference of the page. In the upper right quadrant 4 positions of the rod are shown with respect to their momentarily co-moving frames.

detail by Rindler[9] and is sometimes called Rindler motion. The points at unit spacelike interval from the origin of a Lorentz coordinate system in 2-dimensional spacetime form a hyperbola, $R = 1$, as in the figure above (we need consider only one lobe). So do the points at 2 units spacelike interval, $R = 2$. These hyperbolae are asymptotic to the light cone in that quadrant of spacetime. Imagine these hyperbolae as endpoints of an accelerating unit-length rod. Any spacelike line through the origin that intersects the first hyperbola will intersect the second and each segment so defined will be both a unit spacelike length, and a unit *spatial* length with respect to each momentarily co-moving inertial frame. Let every other point in the rod have a corresponding hyperbolic acceleration, constant in its interval from

---

[9]For a deeper discussion see Rindler (2001:§3.8); also Misner et al. (1973:§6.1, §6.2); Schutz (1985:56,150) where it is the subject of student exercises. Lyle (2010b) gives a careful, perceptive analysis of the problem of a rod accelerated as rigid, including not only the kinematics of accelerating points, but also the problem of what forces could accelerate a rod in this way. His viewpoint differs from that of this chapter. There is an excellent recent discussion of the rockets and thread paradox: Fernflores (2011).

the origin. The equations of motion of the end points are:

$$t^2 - x^2 = 1^2 \qquad t^2 - x^2 = 2^2$$

The equation of motion of each rod-point at $R(1 < R < 2)$ will be

$$t^2 - x^2 = R^2$$

In each momentarily co-moving inertial frame the rod has unit length; the x-coordinates of points in the rod (in these frames) describe acceleration fields that give the geometric structure of the rod and are the *same* in each frame. The rod at any moment is Lorentz contracted relative any earlier frame as a function of its velocity in that frame. Conversely, it is slowed and expanded relative to any later one. At each spacetime point on the worldline of a rod-point, the acceleration 4-vector is the same in magnitude and orthogonal to the velocity 4-vector at the point. The directions and magnitudes of the acceleration 4-vectors attached to points on different rods vary continuously with R. It is not identical acceleration.

In the lower right quadrant of the figure the rod is shown as slowing and expanding relative to the frame defined by the t and x axes of the coordinate system shown. In the upper quadrant the rod is shown in each momentarily co-moving inertial frame.

The example may be generalised by appropriate changes in the magnitude of the constant acceleration vectors i.e. the curvatures of the various worldlines for each thread-point. In the figure, curves closer to the origin are accelerated more than those further from it. A thread, accelerated according to this pattern, will not break. That is so whether the acceleration is gentle or not.

However, the example is idealised. It assumes an unrealistic continuous, homogeneous rod or thread. To characterise force fields that could produce the pattern for any particular rod would be enormously difficult. In real rods, quantum structures forbid continuous matter. Further, the forces would vary from rod to rod – different materials present different dynamical problems. Even in rather homogeneous cases there are different internal micro-variations of mass. Here, at last, is a serious issue of dynamics that depends on how the rod is made up. However, such dynamical complications are beside the point of the lesson taught by uniform acceleration. What informs us about Lorentz contraction/expansion is simply the *kinematical* pattern of non-uniform motion for each point. That description draws only on the geometric structure of the 4-object's motion and on the spatio-temporal environment in which the object moves.

Thus, in Minkowski's ontology and especially in the kinematics and geometry of its objects and their spatial-temporal environment, we find a limpid constructive explanation in purely kinematic concepts of why a moving rod contracts in SR. We find, too, an explanation of why this spacetime theory has Lorentz invariances and covariances. That is the glory of the entity that is Minkowski spacetime.

I conclude that Bell's paper, while ambiguous, is, in none of its possible senses, a correct teaching of SR. If the dynamical interpretation rests on Bell, it, too, is ambiguous and incorrect. But it is unclear how far it does rest on Bell. In any case its explanation terminates with the brute fact assertion that all fields are Lorentz covariant whereas Minkowski's interpretation explains this clearly, deeply and comprehensively.

6 TIME AND SPACETIME: THE SAME ONTIC TYPE

> Chapter 6 gives a partial answer to the question how time can be part of spacetime by dismantling the claim that a flow of time is essential to it. One of the ways in which SR is puzzling is that it looks flatly inconsistent with flow. Its time looks much more like space. I argue outside the context of SR that time, as naively but correctly conceived, neither flows nor is static. We are inclined to think that it must be one or the other because we liken it too closely to space. My argument does not rest on anything that could be called a theory of time but only on principles of semantics and metaphysics that are in common use in other areas. I conclude by suggesting how models and diagrams can properly represent time without giving rise to confusions about its flow.

6.1 Introduction

This chapter considers time by itself outside of SR in order to explore how it can be of the same ontic type as spacetime. Obviously if there is spacetime there is not also a separate entity, time. That does not mean that they are somehow antithetical although, on the face of it, they do seem very different. I give a light treatment of some purely philosophical problems that impede our understanding of its type. I make out my own view on this by contrasting it with others, but I do not attempt a full critique of opposing theories. More complete or contrary accounts may be found in Broad (1938:Chapter XXV), Prior (1968), McCall (1994:Chapters 1, 2), Tooley (1997:Part One), Mellor (1998), Maudlin (2007:Chapter 4), Dainton (2010:first 7 chapters). The next chapter is about how time works in spacetime. Because it already incorporates time, spacetime can't be said to change: there is no time dimension outside it in which it could conceivably change. This is apt to produce the kind of paralysis of mind that Meno complains of when he likens arguing with Socrates to an encounter with a stingray.[1] My present aim is simply to do what I can to release this kind of cramp. In particular, the worry that time either flows or becomes frozen in spacetime evaporates. It is more like unlearning something than any voyage of discovery. Then we can move onto probing the Clock Paradox of SR.

Time ought to be easy to understand, but it isn't. Time is just a single dimension in which events happen and that is just about all that you can *reliably* say about it without dipping into some serious physics. Compare space: space is a 3 dimensional arena in which things and events are related to each other. The physics of spacetime doesn't matter for this chapter. It is about naïve ideas of time. But even here, there is an imperious urge to say more, to add that time *does* something. But how time *could* do anything is more than a bit of a mystery.

We don't well understand what gnaws at us about time, although something certainly does. Thinking about whether we should admit a real Present and a Flow of Time, it is easy to baffle ourselves. I aim to show that it is tempting but dangerous to theorise about time by means of diagrams.

To see how, let's start with couple of theories – with pictures – by admirable philosophers. Their theories turn out to be too complex to portray naïve time. Each suggests a picture, and even entails it in the weak sense that the theory and the picture stand or fall together. The images bewitch us, but they can be probed so as to unmask the structure of their theories. Each reflects a metaphor about what time does – it flows. Metaphors bewitch, too: time doesn't flow. I next go on to look at admirably authored views called presentism. They might be seen as responses to the problem posed by the first two theories (or their ilk). It, too, has pictures and, again, they are wrongly structured to portray naive time and the temporal world.

I will try to talk helpfully about time without developing a theory of it. A picture can certainly go along with this modest approach. It is an obvious one, although seldom considered in the literature. I can see no way of objecting to it. It comes at the end of the chapter.

6.2 Time diagrams

Consider the ingenious theory of Storrs McCall (1994). It is candid, direct and, almost everywhere, limpidly written. From one perspective McCall's work is about the metaphysics of quantum theory and less about relativity. But there is also a core message about naive time. This is sketched in diagrams that make vivid that theory's problem

---

[1] Plato, *Meno* 80c

with time. In the background lurks my guess that all theories of naive time stand or fall with diagrams and that diagrams are intimately connected with their ontological troubles. What puts McCall first is that his pictures are so graphic, his theory so elegant.

In McCall's model, the universe is a branched spacetime. It resembles an earlier Everett (1953) many-worlds model for quantum mechanics, but with a twist. First, here is an Everett diagram.

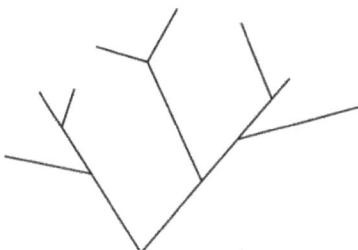

Figure 6.1: Everett diagram

In this, the branches above any time-point are later possible spacetime worlds accessible at that time, and, of course, later than it. The diagram suggests the range of possibilities that quantum indeterminacy leaves open. At any node, nothing marks out any one later branch as preferred. So there is no unique future path through the tree. But, at each point on the tree, a past path is uniquely fixed.

McCall claims that this pictures a *static* world. Time must *flow* if the diagram is to represent the world as dynamic. So he imposes a temporal process on it: the branches successively fall leaving a linear trunk, branched only later than (i.e. above) the point of first branching. The left tree shows the whole of temporal reality at the date of its lowest node. The right tree has lost branches and shows all of temporal reality at a later date. The world-tree undergoes a branch-dropping *process*.

Figure 6.2: McCall diagrams (McCall, 1994:3-4)

The present is the lowest node of branching. The past is the linear trunk; the future is the fan of branching possible worlds. I will repeat these diagrams and pair others with them using the McCall figures to show which new diagram illustrates earlier and later states of temporal reality.

In the picture, time does something. The present prunes branches as it flows upward. The *picture*, or rather the series of them, is certainly dynamic.

This is indeed vivid, but too rich to map naive time. On the face of it, it shows a 5 dimensional spacetime. In an Everett diagram, each point maps three spatial dimensions, and each bit of rising line pictures a time through which the 3-space endures (persists etc.). That's already four dimensions. McCall's diagrams add time (the time in which a movie would roll that shows the branches falling) to make the thing dynamic (McCall, 1994:4, 10). But this time surely represents something and doesn't this have to be *itself* – time *all over again*? Unless the movie rolls (in time, how else?) the branches don't fall. But 5D spacetime is not what McCall means: he claims that only one time is represented. So the question changes: why does time need to be *mapped* twice? In particular, why does it need to be represented *in space* as well as in itself? How does the diagram square with the world intended?

Michael Tooley's (1997) "growing block" universe also has too many time dimensions, quite like McCall's. Tooley's work is replete with fine and interesting distinctions and is much concerned with relations between cause and time. It, too, admits a diagram of naïve time, although Tooley does not draw it. It replaces McCall's future branched possibilities, with a growing past and an empty future. In both theories, the past is "real" at every present. So a Tooley diagram is also a movie and inherits the structural mismatch so puzzling in McCall. Tooley's "at a time" and "as of a time" distinction is no more limpid than McCall in addressing the objection that the diagram includes time twice over and so can't represent naive time. (These theories are searchingly discussed in Dainton (2010:Chapters 6.1-6.5).

Back to the Everett diagram. Why think that it represents a static world? No reason, unless you put it in a movie, too, one in which every still is the same as every other and each represents all of temporal reality. Pair the same Everett diagram with successive McCall diagrams as shown in Fig. 6.4, and Everett doesn't change as McCall does. We have a movie with the same diagram in every frame. It's a movie about something static – a static *diagram* (Savitt, 2006). But any Everett diagram already represents a dynamic *world* in spacetime. Special relativity is a good theory of dynamics, after all, and Minkowski diagrams go hand in hand with it. Why foist the movie on Everett or Minkowski? Especially since it remains obscure how McCall would have us understand how the movie maps a 4D world.

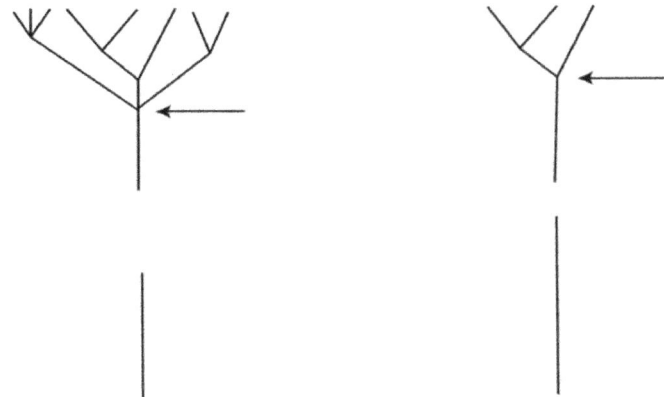

Figure 6.3: Tooley diagrams paired below McCall's

Of course, it's good to have a diagram that doesn't change while you look at it. So lines are generally good in diagrams. But, for picturing time, they can be deeply confusing.

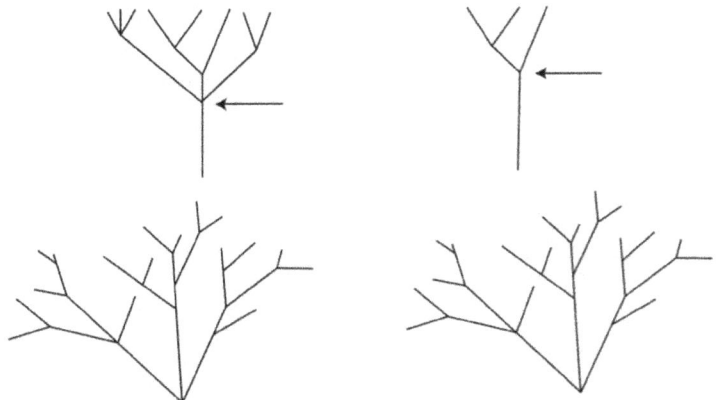

Figure 6.4: Earlier and later Everett diagrams each represents a spacetime world. Since each is the same diagram for McCall-earlier and later stages, the series represents a static sequence of spacetimes

Yet movie diagrams don't do enough. First, what tells us that we mustn't run the movie backwards? That would show a growing McCall tree or a shrinking Tooley block (Dainton, 2010:Chapter 6). More explicitly, the ordering of events envisaged within the (time representing) line can't dictate the order in movie time in which the branches fall or the block grows. Or, to put the point directly from our diagrams, what tells us that we should not read them right to left? Nothing about how the branches are mapped in a still tells us whether the movie should show many branched trees before or after fewer branched ones. There's another difficulty: each still shows all past times at the same (movie) time as it shows the present (the lowest node in the still). It shows events of 12 noon 15 March 1897 as occurring low down in the trunk just when the still with the lowest node at 12 noon 15 March 1997 is shown, despite the former event's being past in 1997. So, right now, there are past people portrayed in 1897 firmly but mistakenly believing that their actions and thoughts are present. See Dainton (2010:Chapter 6.6); Bourne (2006), Forrest (2004). There's surely something wrong with that.

6.3 Temporal solipsisms

Presentism is hard to grasp. Partly that's because of the way presentists often talk about it, partly because it seems to mean something incredible. It's easy to agree that dinosaurs don't exist because their lives are past and that anything that exists is somewhere or other now.[2] But, that sentence, when tensed, is just a truism, so that's not presentism. To say what presentism is in a way that gives us something worth quarrelling over, we have to avoid using tensed sentences and familiar English. If we use quantifiers without restricting their domains to any date or time and delete present tense from our formal sentences, then presentists believe that there are no dinosaurs. By contrast, I believe that I can refer directly to dinosaurs just as I can to cows.[3] Back in familiar English, both of us agree that there were some dinosaurs (See, e.g. Crisp (2003)).

Rea (2003) distinguishes common sense presentism from something else. But I find it hard to come at presentism save from a couple of pictures, neither of which makes final sense to me. Let me prepare the ground for them.

There are two kinds of temporal solipsism. Each is a presentism and they share the view that, in some fashion, only the present is real and other times do not exist. In the one version, there is only one time – the present. This does not mean that it is *always* one and the same time, since, within this form of presentism, "always" can't be interpreted as a quantifier over distinct times. So, on this theory, there can be no other time but one.

In the other presentism, different times may have the property of presentness – at different times, of course – and when they have it no other time has it so it doesn't exist. Call the first *pure* temporal solip-

---

[2] Let's draw a veil over the whereabouts of numbers and the like.
[3] I.e. when I say "Newton wrote the Principia" the sentence is true without the proper name's being nested in an implicit opaque operator or similar construction.

sism or pure presentism and the second *transient* temporal solipsism or transient presentism. For instance, in 2000, so transient presentism runs, only events in 2000 were real and earlier and later events just didn't exist and weren't real. But now, only 2013 events are present and real and those of 2000, like any other past events have no reality at all. Nor do future events have any.

Thus each view treats not only the future but also the past in the same way as Tooley treats the future. At any time, only that time is present, and other times, past or future, are not real *then*. The last sentence holds even if the quantifier "any" ranges over only the unit domain of the one and only time.

Prior (1970), Bigelow and Pargetter (1989) and Bigelow (1996) are pure presentists – certainly Bigelow is, and I believe the same is true of Prior although I think he did not explicitly make the distinction just drawn.

On either view, what is to be made of true statements ostensibly about the past? Prior treats "Past" as a sentence-forming operator on sentences with a core function of redirecting the semantics of the sentence operated on. For instance the sentence:

Past [A battle is fought at Waterloo]

says that a battle was fought at Waterloo. It seems to assign the contained sentence to another time analogously to the way a modal operator assigns a sentence operated on to another world. Some of Prior's analogies strongly suggest redirection to another area of *discourse* – "in fairy stories..." or "in Greek Mythology..." He explicitly rejects the analogy of redirection to anything like other (real) regions. Other times are not other *regions* of time. This sounds like cancelling the semantics of the contained sentences.

Bigelow prefers to construe sentences containing "past" as propositional functions, as complex properties true of present reality. Within presentism, I prefer that, too. He rightly rejects the view that, e.g., truth-makers for sentences about the past, are just present *traces* of past events. Truth makers for past tensed propositions remain real but irreducible *properties* of the real, that is, the present, world. So there is past and future in that there are subtle properties – "having happened" properties, subtle in being spatially unlocalised – true at the one and only time, which yet have some kind of time-directed sense. They are like tangent vectors at spacetime points, representing directions at points without themselves being extended in spacetime in that direction.

Solipsists have a diagram, too. But what is it? Unlike an Everett map, the diagram is just a point, representing three spatial dimensions. Nothing else exists to be represented.

•

Figure 6.5: The only real time is represented by an unextended dot

That is Bigelow's choice.[4] However, while we need not accept McCall-like claims that it is a static world, it is, nevertheless, a changeless one. Nothing happens, or can happen, whatever subtle properties are built into the point. The properties only mimic real change. In Prior's version, there is talk of change, but it is all talk.

Transient presentism has a different diagram, and I suspect it plays a hidden role in how presentism is generally understood when it seems at least minimally plausible. Or, more modestly, if it *doesn't* play this hidden role, then I cannot grasp what presentists mean. I should add that Bigelow disclaims any argument from plausibility, aiming just to present pure presentism as consistent. However, he says that he has now come to think it believable.

Here are two transient presentist diagrams:

Figure 6.6: Earlier and later presents are pictured as temporally unextended but qualitatively different

Under a McCall still, draw a dot. The dot represents a 3D world. Under the "next" McCall still, draw another dot, but of a different colour (shape). What does this mean? It represents the fact that different present-tensed propositions are true in the different worlds, so that the dots represent different past and future "vectorial" properties. So there is change, even though presentist time is not diagrammed by a spatial extension. If we accept the movie representation, then the present changes its qualities in whatever dimension it is that movie-time represents. On the other hand, if the diagram is the same for each

---

[4]If I understood him correctly in conversation.

pairing, the world is static, unchanging in that sense. Only something like colour, a quality of the dot, not another dimension, can picture presentism's change on pain of clearly destroying the 3-dimensionality of the presentist's world. Colour has become another representing device, picturing change. But, again, colour is a less obvious fifth representing dimensions. It doesn't represent a real past since it flouts the convention of representing that by spatial or temporal extension. Change is assigned to qualitative differences.

6.4 Existence, occurrence, truth, reality

Famously, Augustine said that he understood time well enough while no one asked him to say what it was. If they did, he was stumped. We might take this as advice. Stop thinking about what time really is, and more particularly, about what it does. Think about other things instead, and many difficulties vanish. On the one hand, good theories of existence and occurrence, truth and reality bid fair to erase several puzzles without ever broaching the topic of time itself. On the other hand, theories of temporal processes, especially the phenomenology of experience and action, may solve many others, and theories of temporal processes are not, in general, theories of time.[5]

Augustine didn't take his remark as advice, and the result was brilliant but unlucky. It may seem that I hold, tacitly, a theory generally described as the B-theory of time; that's a doctrine that denies any special ontic or other distinction between past, present and future time. But if it is only that *negative* doctrine, it is not properly a theory of time, I shall argue. McCall, Tooley, Prior and Bigelow all argue for theories of time that are certainly inconsistent with the doctrines that make up the B-theory. They posit ontic claims about the structure of naive time. They really are theories of time and, I have argued, false theories. But the negation of a theory need not be a theory and the so-called B-theory is (or it can be) just a conjunction of principles derived from and motivated by wholly non-temporal metaphysics. I go on to sketch principles governing truth, existence, occurrence and reality. They go together in some fashion. What I have to say about them has been well said before, so I sketch familiar arguments briefly just to illustrate that none of them is motivated by puzzles about time. They say nothing about it. Yet when these morals are applied to time in the same spirit as one applies them to anything else, some theories of time are ruled out - but not by an opposing *theory of time*.

The first word about truth (not the last, certainly) was said by Aristotle. Briefly, "p is true" adds nothing to the content of p. It

---

[5] A theory of the evolution of hurricanes, for instance, is not a theory of time. However, theories of some processes do lead to theories about time, as SR did.

doesn't tell us that p is known, for instance, let alone how or how well it is known. It says nothing about a speaker's frame of mind, not his beliefs, his confidence, his wish to dominate or bully[6] or anything else.

In particular, while a good Aristotelian may ask when p's truth-maker occurs, or when events mentioned or entailed by p occur, "p is true" admits no question or addition as to when it is true. It is as absurd to speak about when p is true as to speak about where it is true. "Italy is shaped like a boot" is some sort of spatial statement that has its truth maker in Italy. But that doesn't mean that Italy is the place where it is true. That the statement is true wherever you make it doesn't mean that it is true everywhere. Statements are not true *at places*. Nor at times: not any time nor all the time, and for entirely parallel reasons. Truth values are not eternal. These arguments are no defence of anything that can properly be called eternalism.

Another principle. Respect the distinction that *things exist* whereas *events occur*. It is very apt to confuse if we say that things occur and events exist – especially the latter. The 2011 cricket test match in Adelaide doesn't exist, and didn't exist; of course it did occur and was, indeed, played.

Next, the following expressions don't describe. "Tigers exist" says the same as "There are tigers" says, which is not the same as what "Tigers are there" says, for that indicates a location. If there are tigers we can always ask where they are, because tigers, unlike, for instance, numbers, are understood as spatial things. Existing tigers are not a kind of tigers, not even a spatial kind of them; as Austin said, existing is not like breathing only quieter. "Thunderstorms occur" says what "There are thunderstorms sometimes" says and, whereas "exists" never happily applies to events and processes; "occurs" is explicitly about events, properly only about events and about their being at times. Nevertheless events that occur are not a kind of events; not like explosive events only quieter.

The trouble with "exist" lies in a hint at endurance. Thus "The past exists" already suggests that it endures, just as in the theories of McCall and Tooley. But "exists" does no *more* than hint at that. If numbers exist, they don't endure. Events don't endure.

"Real" is like "exists" and "occurs": real things are not a kind of things. Unlike the other two, however, "real" takes on something like descriptive implications, given a background sense of "unreal," and relative to a kind. Toy cars are not real *cars*: you can't seat people in them and drive them on the road; but they are real *toys*. "Real" may be applied in a somewhat different, metalinguistic sense to contrast terms which do refer with those which don't. Gnomes are not real,

---

[6]I have in mind Nietzsche's misunderstanding of truth and the confusions that some postmodern styles of thought have drawn from it.

but dwarves are. That is not a difference in kind. Reality is not a property that dwarves have and gnomes lack.

Next, these principles strongly suggest that tensed sentences are used indexically. An indexical expression must be related to its occurrence in an act of speech to fix what it refers to. So "I" in my mouth on an occasion of speaking and in your mouth on another refer to different people. They fall into a quite special semantic category: their use in language is indispensable in practice, yet their semantic analysis shows that each is true only in virtue of non-indexical expressions. "I breathe" is true in that the writer of that instance of the sentence breathed when it was written (Mellor (1998:Chapter 3), Dyke (2007)).That does not imply a *translation* of tensed language into tense-free language. Tenses, especially, have to be indexical, since the only other kind of message they could convey is something improper, about degrees or kinds of reality. "Past" can't mean or imply "Had reality but lost it." In this, time and space sentences are semantically analogous. Different times don't give rise to different forms of reality, just as different places don't. They don't, not mainly because time is like space ontologically, but because it's nonsense to suppose that they (or anything else) do or can differ in form of reality.

Let me repeat: these Russell-Quine-Austin doctrines are a theory of existence, occurrence etc. and so debatable; but they aren't a theory of time. Nevertheless, they oblige (at the very least, incline) tenses to have an indexical foundation. What I need from these theories can be captured in one principle of reference, R, illustrated in this example:

R: When I say "Plato wrote Meno" I refer directly to Plato

Tactically, you talk easy-to-follow sense about time by sticking to carefully tensed language whenever you can. Indeed it is fairly hard[7] to talk nonsense about time in tensed language. "Events have happened before now" is far better than the tense-garbled "The past is real." No one is likely to cavil at it or muddle the message. So for "Later, some events will happen" as against "The future is real." Again, better to say "The only events that are occurring are present ones" than "Only the present is real." You can even defy Augustine's caution and say what time is: time is that arena in which past events have happened, that in which present events are now happening and that in which future events will happen. Boring, maybe. True, certainly. Not easily misunderstood, crucially.

Thus tensed talk has a big advantage not just in contexts of practice but also to clarify the point we try to get our tongues around when we ask metaphysical questions about past present and future happenings. To say that, at any time, there are events earlier than those of the time

---

[7] Alas, not impossible.

and later ones as well, lacks the aura of profundity of a question about how the Battle of Waterloo stands, ontically, *now*. But it gains in clarity. The most useful answer to that question – it really happened – is perfectly explicit and satisfactory, provided one recognises the semantic facts about indexical tense.

While you might avoid pitfalls talking tenseless language, surprisingly many very bright people don't avoid them and I advise tensing whenever you can. It's even dangerous to disagree with theories of time in tenseless language. The denial is apt to inherit the bad sense denied and is even more likely to be so interpreted. Witness the idea that in the so-called B theory of time, time is static, that it portrays a block universe or that it can be illuminatingly dubbed eternalism. But you can't always avoid tenseless language. Rule R has to be tenseless to do the job required of it, but at least it is only a rule and not about concrete happenings or about time itself.

There are also "natural" examples of tenseless speech. "It never rains on the moon" quantifies over all times and resists a tensed reading. Some will insist that it be read "It never has rained on the moon, isn't raining now and never will rain" and might claim that this is different from the tenseless reading. Not if tenses are indexicals, however. But the tenseless reading is quite natural – if only people will agree so to read it!

6.5 A naïve model of naïve time

Why think that diagrams are the problem? Because the purely spatial basic structure of diagrams is part of what tempts us to tack on another dimension. (But serious confusions about time itself play a main role, too.) Diagrams are good for representing space, because they are just *bits* of space and share only spatial characteristics with space. A quite different feature of spatial diagrams is useful: we can *go on* looking at one picture of a single moment (as in a McCall still). But that can be confusing. However, we are not confined to purely spatial diagrams of naive time. McCall and Tooley diagrams resort to movies, for instance. Once we allow that time might somehow represent itself, there is a very direct and simple kind of diagram of time and of spacetime. Simply use a *bit* of time or, better, of spacetime.

I suggest three diagrams all based on the same principle. The first is an hourglass, the second a clock, but for fun and vividness, the novelty toy called a snowstorm does the job well. In each case the whole diagram is the hourglass, clock or snowstorm *in process*. In each case, the result is a kind of movie. But now, a bit of time pictures time *and nothing else does*.

Notice three features of these diagrams. First, each has 4 dimensions and so as many representing as represented dimensions, since each dimension represents itself. Second, each diagram (model) plainly represents a dynamic, not a static, world. Thirdly, and oddly perhaps, none of them tells us anything we seem to want to know about naive time. That, I claim, is because it is not the diagram of any *theory* of time, so it is trivially but exactly right. Perhaps it fits what is called the B-theory, at least in the sense of being consistent with it. But only if that is not a theory of time. Or, if it is framed as a theory, it is not what I advocate. We need not say that the active diagram has nothing to show us, just that it displays only what's obvious.

It nowhere shows what McCall or Tooley want – the falling of future possibilities or the adding of slices to reality. Nor what Prior and Bigelow want – that *now* is all. Nor does it show any other of the above. Yet if they're right, what they claim must be fully modelled in a good diagram.

Is this an argument that what they want in their diagrams is not there in the world? Maybe! Although the time extended (dynamic) diagram is trivial, it can hardly be mistaken. That isn't why we are disappointed with it. What draws us to go beyond it looks like the very confusion we are trying to identify and eliminate.

Yet perhaps it's no objection to these theories that we don't actually see what they should claim is in the diagram. After all, if the Everett many-worlds theory were correct as quantum physics, our diagram wouldn't portray in time what Everett diagrams portrays spatially – the branching, that is. Yet Everett diagrams do aim to portray a feature of the world we would like to grasp – quantum indeterminacy. Well, if we do live in an Everett branching world, (I doubt it) my diagram branches too. At equally long intervals after the first shake of the snowstorm toy, there are, elsewhere or in other worlds, counterparts of me watching the different swirlings and settlings of the "flakes" going on in each of them. That we don't see this in our branch of spacetime surely doesn't mean it isn't obscurely there.

However, the Everett diagram branching hypothesis *doesn't spring from anything obvious*. What motivates it are puzzling and not at all blatant quantum difficulties, so we shouldn't expect to see the branch switches. By contrast, philosophical theories of naive time all spring from the sense that something obvious needs to be metaphysically accounted for.

Not something merely *a bit* obvious, either. Unmistakably obvious – glaring! The impulse to star the present springs from an impulse not just deep in the way we experience what happens, but obvious beyond any possibility of our missing it. The impulse cannot rest on any inference, since someone might then be too stupid to draw the conclusion. Impossible! Nor do we think that even the most careless observer could overlook what pervades simply every experience, immediately, vividly, unmistakably. If it is there in the diagram or in our experience then it ought to be inescapable, utterly beyond us to fail to find it.

But the falling of branches from spacetime? The adding of layers of reality at each moment? The ontic failure of every time but one? There is nothing immediate, either visible in our short movie, or in our experiencing of it, that corresponds to these arcane suggestions. They are just too remote from experience to be what we are after. Surely they are wrong answers and probably to wrong questions. We've looked in the wrong place. Time certainly goes on puzzling me. But I doubt that a philosophical theory of time – let alone another diagram – is the right place to look for answers.

The snowstorm diagram is right; the other diagrams are just plain wrong!

All this suggests that the solution to our problems lies in a sound theory of the phenomenology of time and especially, I think, of action. Some interesting work has been done in this direction. (See Butterfield (1984), Nerlich (1998), Dainton (2006), Paul (2010)). But, so strong is the feeling that the present is The Time, that it is hard to believe that a mere theory of *any* kind can do the job of showing that it is an illusion. My phrasing here is deliberately loose and vague since even the attempt to characterise what "present chauvinism" is (I got the last phrase from Heather Dyke, who may well have coined it) risks begging serious questions. Perhaps time perception is the product of a modular structure in the brain so that an inarticulate sense of passage is thrust on us, probably for good survival reasons. But in that case, we may never solve the problem of how we experience time.

6.6 Time, space and spacetime: the same in ontic type

The main point of the chapter is that time is ontically like space and spacetime in this respect: there are no ontic distinctions among times just as there are none among places. I really did eat breakfast yesterday and I really will die some day. That doesn't mean that time is space. Some may feel that it does, but SR represents the two quite differently. The signature of the metric has different signs for space and time differentials. We don't well understand time even now. Time is directed and space is not. It isn't clear what that amounts to. A solid body of opinion puts the direction of time back onto the directedness of its contents. There is the 2nd law of thermodynamics: entropy increases with time. I cautiously incline towards that view. But it does not make time itself different from space.

7 How the Twins do it: SR and the Clock Paradox

Chapter 7 has two aims: it expounds the notorious twins or clock paradox of SR in rather elementary fashion and it shows how the conception of time in chapter 6 works to explain this feature of SR. This explanation of "the slowing of a moving clock" is compared and contrasted with another according to which time flows, and flows at different rates in different frames of reference. The former explanation is preferred. At this stage it should be clear how time and space can both belong to the same ontic type as spacetime.

7.1 Introduction

In an interesting and controversial paper[1] McCall (2003) (hereafter "the authors") claim that the notorious twins (or clock) paradox in SR may be understood not only in the context of 4D geometry but also, and equally well, in a 3D setting. They claim that 3D and 4D descriptions are equivalent in the sense of being inter-translatable within SR (§4). Their 3D analysis aims to solve the twins paradox through the use of A-theory concepts and, in particular, by using the idea that time flows at different rates with respect to different frames of reference (115).

We can put the idea of time in the last chapter through some paces by using it to shed light on this notorious problem in SR.

I will assume (as the authors do) the thesis of 3D/4D equivalence in the sense that 3D and 4D descriptions are inter-translatable. Under that assumption, I study the paradox in terms of 3D/B-theoretic concepts, a style left untouched by the authors. My solution calls on

---

[1] Numerals in round brackets refer to pages in this paper.

the proper times of the twins (or clocks) and makes no use of different frames of reference. This implies that neither the A-theory nor alternative frames play a necessary role.

7.2 What is the twins paradox?

The twins or clock paradox is perhaps the best-known consequence of SR. It is a paradox only in that it clashes strongly with the intuitions we have about time before we are at home with the theory. Although the paradox is familiar, I'll sketch it and its 4D solution.

Let there be two ideal[2] clocks which both read 0 at some point-event P. Thus they agree in reading 0 together at the same time and place and so they are absolutely synchronized. After various vicissitudes the clocks meet again at some later point-event Q and their readings are compared again at that place and time (hence, again, the comparison is absolute). In general, the readings on the clocks at Q will disagree. We may add colour to the example by substituting for the clocks a pair of twins.[3] The twins (Jack and Jill) leave P at the same age, but when they meet at Q one will be younger than the other (in general), depending on how they have journeyed.

It's customary to use a simple but special case to illustrate the general claim. Jack remains at rest in an inertial frame throughout and his sister Jill moves straight out and straight back to a spatially distant point R (far and fast, let's suppose) on her journey. Let Jack's inertial frame be the one with respect to which P and Q occur at the same place but at different times. Jack's age (hence his clock's time) is a (maximal) limit of all the ages (times) of twins and clocks that are or could be at both P and Q.

The authors' 4D explanation uses a sketch of the space-time trajectories of Jack and Jill in the simple case. (See Figure 7.1) In fact, the figure can equally represent something simpler still, the case of three inertial clocks, two of which closely approximate Jill's space-time trajectory. Their temporal lengths also closely approximate the interval measured by Jill's clock. Although the figure makes an angle in spacetime at R, rather than a curve when Jill accelerates to start back again, what happens just in that period is largely irrelevant to explaining the different ages of the twins, as the authors agree (119). The diagram is also helpful in that we see at once that the twins' journeys are not symmetrical in space-time: the direct path from P to Q

---

[2]SR admits the concept of an ideal clock, one for which accelerating forces do not affect its measuring of intervals along its curved worldline. That is the clock hypothesis. See Rindler (1977:43).

[3]Clocks provide the advantage of letting us toy with numerical measures of time in describing the paradox.

does not equal the dogleg path that goes via R.[4] Thus the clock on the direct path measures a different space-time interval from the one that goes via R. It does not measure it in a different way or by a clock that runs at a different rate.[5]

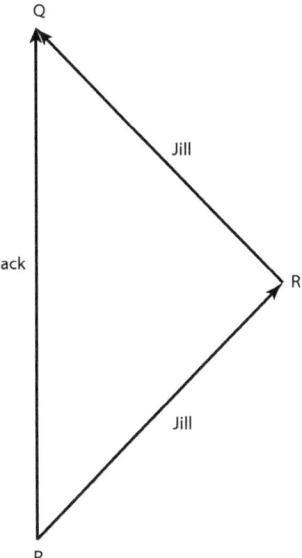

Figure 7.1: This illustrates the simple case where Jack is at rest in an inertial frame and Jill's worldline is PRQ

In the general (rather than just the simple) case, neither Jack nor Jill need remain at rest at any stage in any frame of reference. Their journeys are arbitrary timelike curves from P to Q. Unless their journeys are the same temporal length from P to Q, their ages differ at Q.

7.3 3D description and the A-theory

The 4D explanation makes no use of coordinate systems or frames of reference. The age difference is frame invariant, as the authors agree (115, 121). It is not obvious why translating the explanation into 3D language should – or how it could – exploit different frames of reference. Of course a 3D description obliges us to think of space and time as somehow separated, but why do we now need different frames of reference to explain what we explained before without them?

---

[4]The Euclidean diagram misrepresents the Minkowski geometry in one respect; longer Euclidean lines correspond to shorter temporal lengths in Minkowski's metric.

[5]Taylor and Wheeler (1992:76-77) are emphatic about this.

The authors do not discuss what rules of translation underlie the claim that 3D and 4D descriptions are equivalent (in the sense of inter-translatable) within SR. My translation of the 4D picture of the simple case is different from theirs, so I should defend it. But otherwise I won't consider what they say in introducing the A-theory into the context of the twins.

The authors translate "temporal length" in the 4D description of the simple case by the words "process" and "rate" (121). They claim that these words are conceptually bound to 3D language, and, more contentiously, bound to the much richer A-theory. I doubt both claims, but consider only the second. I note, first, that 3D description is not necessarily A-theoretic. Further, a 3D/B-theoretic definition may be given of "process" and "rate": a process is a set of states of some continuant, differing at different (earlier or later) times; a rate is a measure of how many changes per unit time occur in a process. Nothing in this implies a real (i.e. not merely indexical) present or a flow of time. Finally, a translation of temporal length (as a 4D concept) into temporal rate as a 3D one (121) is neither apt nor necessary. Temporal distance in 4D language and "process" and "rate" in 3D language are not "two sides of the same coin" (121) because neither of the latter terms refers, as "temporal distance" does, to quantities of time. Instead, "temporal length" and "temporally long" translate just into "long time." To say that one journey takes a longer time than another is the commonest of common (and hence 3D) language. Of course, in saying that, we usually assume that the journeys either began at different times or ended at different ones. But in the paradox, where the journeys begin at the same time and end at the same time, the phrase still tells the story accurately. That is exactly what jolts our intuitions.

This does little to explain the paradox. It just restates it in a particular, accurate way.

Clocks don't measure "rates of elapsed time" or measure "rates of temporal flow" (121). They measure time: what time it is and that it's a long time since lunch. "Rates of time" is unclear, calling for special justifying argument and explication.

7.4 A 3D B theory explanation of the paradox

The paradox is not unique to SR. Lorentz (1904:§4) deduced a virtual equivalent of the slowing of a moving clock. Its spatial counterpart, the Fitzgerald length contraction, was also prominent in the literature before 1905. The idea of frames of reference is alien to these early findings. Motion is absolute. Let me begin at something of a tangent, then, and deduce the twins paradox in a simplified (ether free) Lorentzian absolute space and time theory. It is easy. Then we need to concern ourselves with only the metaphysical differences between this theory and SR, ignoring the physics of the ether. Simplified Lorentz theory has attracted a number of philosophers (e.g. Prior (1970), Tooley (1997), Craig (2001)).

In this theory, as in SR, it is a fundamental principle that the c in Maxwell's equations for the electromagnetic field is a finite constant and has the dimensions of a speed. Let us assume that this is the speed of light and electromagnetic radiation.[6] This means that we can construct an excellent clock, a Langevin clock, from a couple of mirrors at a fixed distance from, and parallel to, one another. Between them a photon bounces and a counter counts how many bounces strike one of the mirrors. If we take the constant, finite speed of light as making visible the fundamental speed of the theory then it is fundamental that this makes a good clock.

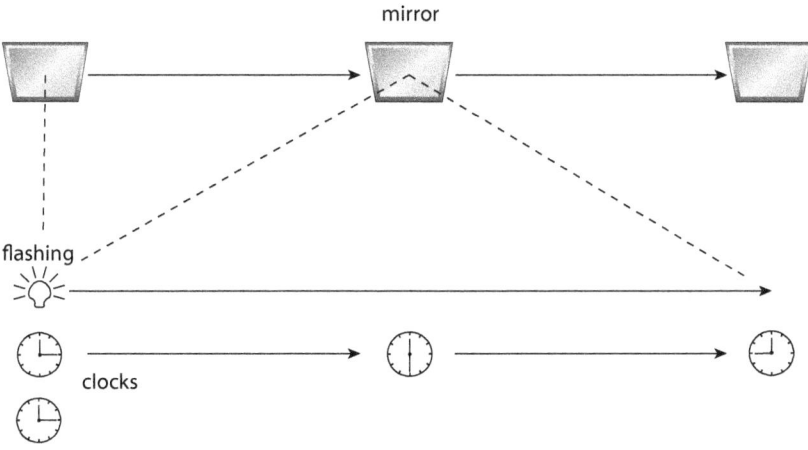

Figure 7.2:

Now suppose that Jack is at absolute rest, so that his clock is the gold standard of absolute time. That is, the bouncing photon that drives his clock moves between the mirrors along the same path up and down. Jill moves off in a straight line at uniform speed, taking her clock with her. For simplicity, suppose that the line of her motion lies in the plane of one of the mirrors, so that it is orthogonal to the

---

[6]This is the customary context for these discussions rather than one discussed in Chapter 3. The assumption could fail if, for instance, the photon is massive. Then we would have to turn to Minkowski's basis for the structure of spacetime. But we may ignore that.

direction of the bouncing photon that works Jack's clock. Then her clock will run slow because the photon will not move up and down along the same track orthogonal to the planes of the mirrors, but along different and longer paths at angles to her mirrors (Figure 7.2). So her photon bounces less often and her clock's counter will tick less often than Jack's. This has nothing to do with how Jack measures Jill's clock. Since her motion is absolute, the paths her photon takes are absolutely longer than the path along which Jack's photon travels. The slowing is absolute. Jill's clock simply doesn't measure The Time.

Clearly enough, if Jill reverses the direction of her journey and returns to Jack, the photon working her clock will still move along slanting paths between her mirrors, so that when she returns, her clock will have ticked less often than Jack's. That will also be true when she accelerates. The only way the path can be minimal (up and down along the same track) and the clock tick properly, is if it is at absolute rest.

Most kinds of clocks involve electromagnetic principles (which underlie the constitution of matter as we ordinarily encounter it), so we can extend the light clock finding to electromagnetic behaviour broadly and thus to ageing processes at large. Jill's wrinkles (and her thought processes) involve them, too. So she will age less than Jack in the same absolute time.

In 3D/B-theoretic language, that explains why Jill is younger than Jack. Her clock and her wrinkles don't reflect The Time. She has aged at a different rate. She has lived through the same period of time, but at less cost in ageing.

7.5 Frames of reference

As a relativity theory, SR tells a different story which departs from the simplified Lorentzian one just in its metaphysics. Let's turn to that.

If we use 3D concepts in SR, then we need ways to separate space and time. If we also want a global story within which we can compare and measure everything in a single perspective, then we need a global frame of reference. We need one to embrace both Jack and far-off Jill in a single view. That entails that we will use coordinate time, the global time of the frame we choose. But coordinate time is not the time that matters: proper time matters. I pursue that theme shortly.

But first, there is an obvious frame for the global coordinate story: the one in which P and Q happen in the same place. In the simple case, that is also the frame in which Jack is at rest. Choose that frame and the global story is just as I told it in Lorentz space-time, save that the photon path lengths are defined relative to the frame, not absolutely. Choose another frame and the story would change in respect of path lengths relative to the new frame, and so would coordinate parts of the description. But not the ages, the age difference, and how many times each clock ticked. They are not coordinate quantities, but proper ones – invariant, as the authors agree. Those quantities are the ones that matter and coordinate times do not, in general, bear directly on them.

Proper quantities emerge if we separate time from space locally for each twin. There is a metaphysical reason for doing this.

Lorentz absolute space and time allow us to define ontologically a state of absolute rest, but there are two problems with it. The first is epistemological and long familiar: there is no way to tell when anything is at absolute rest and when it moves. Measuring instruments "conspire" to hide the difference between rest and uniform, force-free motion.

Worse follows, although it is less often mentioned.[7] There is virtually no access to intrinsic physical properties. Let mass be our example. In the Lorentz world the absolute mass of a moving thing increases as a function of its speed, even when speed is uniform and force free. Mass is an intrinsic quantity, but there is no telling what it is. When we measure the mass of something moving absolutely with us, we get its apparent-rest mass, not its real intrinsic mass. If absolutely at rest, we measure its intrinsic mass correctly, but have no way of knowing that we do. The ontological cost is that intrinsic properties change with mere motion; the epistemic loss is that the world conceals them. Since we can define all classical physical properties (except for charge) in terms of mass, length and time, the epistemic barrier between us and the real world of intrinsic properties is virtually complete.

Thus the Lorentz absolute theory admits inaccessible, intrinsic, inconstant properties that change without the action of causes.

SR sheds these problems: it can allow proper quantities. These correspond to the Lorentzian inaccessible, intrinsic properties that are inconstant without being caused to change. But now they are invariant, accessible properties and inconstant only when caused to change. A thing may change its mass because it is heated, but how could the relative motion of some other thing change its mass?

In SR, if we separate space and time in a local, natural way, without using global frames, then intrinsic properties are accessible. It is only in global coordinate perspectives that quantities vary. In itself, a clock tells the right time, its own time, its proper time. Jill's does so even when Jack says that it accelerates. Jill's clock does just what Jack's clock does: it counts the bounces of the photon between its mirrors and, as she encounters it, the photon's path is straight up and back. Nothing happens to Jill's clock: its intrinsic, proper rate is the same

---

[7] In fact I know of no discussion of the points that follow explicitly as metaphysical ones, although plenty in relativity texts pretty directly implies them.

as Jack's. So Jack takes longer to get from P to Q than Jill takes. Jill is as young as she looks, since she has lived a shorter time than Jack. The proper times are the real times.

Proper quantities in general – proper time, proper mass, proper length[8] and the others – are intrinsic. Proper time, conventionally written "τ," is a clock's own time, proper mass an object's own mass.[9] They are proper not to inertial frames but to local concrete objects (including systems) and processes themselves. They are invariants, but not because the Lorentz transformation juggles them out unchanged. Their invariance is a demand on the transformation because they are the natural intrinsic properties of 3D things once we abandon absolute space and time. Proper mass is identical with rest mass, but we do not need the global frame in which the thing is at rest to define or speak of it. On the contrary, the global perspective can work only if the proper mass – the intrinsic mass – is an invariant property of the body. The relativistic mass, m_i, is a function just of proper mass, m_0, and of the frame velocity, $\mathbf{v}$, relative to a frame, F_i. That's so for our target quantity, proper time, too. The proper time, τ, of an ideal clock (or the proper ageing of a twin) allows the relativistic (coordinate) time, t_i, to be a function just of τ and the velocity of the clock or twin relative to a frame, F_i. Proper time, not tied to one or another frame, is the fundamental concept.

In the simple case, in Figure 7.1, Jack and Jill each begin in free fall (or free float). When Jill's clock ticks 10 times on the outward-bound part of her trip, then the proper time from P to that event is 10 units. But that event is far from the event at which Jack's clock ticks 10. The only way to relate those events is through the global perspective of some frame, and different frames relate them differently. Nevertheless, all agree on the proper times from P – 10 units in each case. What makes the later meeting of the twins significant for a comparison of the real, proper times of each is that the local comparisons obviate the need for any frame comparison. Then we see that Jack's journey took longer than Jill's – that the proper, true, intrinsic times are different. Frames have no role to play in it.

If that is right, then the twins don't do it by having different rates of clocks in different frames, let alone different rates of the flow of time in different frames. They do it by taking different spacetime journeys, one taking longer than the other.

The rates of clocks can be meaningfully compared when both measure the same time interval, as the twins' clocks do in the Lorentz setting. In relativity, the twins' clocks do not measure the same inter-

---

[8]That is a trickier concept, but its complexities can be ignored here.

[9]Proper mass may be intrinsic to a dynamical system of objects, or to an object qua system. It is not as simple as this suggests as §4.3 makes clear. See also Taylor and Wheeler (1992:224-28).

val. We can indeed speak of the interval from P to Q, the length of the timelike line from P to Q. But neither Jack nor Jill measures this interval, save for Jack in the special simple case. Comparison of their clock rates makes no sense.

Spatial journeys with the same departure and arrival points generally differ in length. Odometers can measure such lengths directly. Plainly, nothing about the way odometers measure spatial journeys makes sense of why the tyres on Jack's car have worn more than those on Jill's. Simply, Jack's journey is longer. Analogously, Jack is more wrinkled than Jill because his timelike journey is longer.

In the simple case, Jack's proper time is also the coordinate time of his rest frame. But we can't go on to consider Jill's coordinate time. She has no global frame. (Rather, the *obvious* ways she might try to take a global view yield contradictions.) In the general case, neither need have a natural global perspective and comparing their journeys globally, step by step, is much easier from the perspective of an inertial frame, like the obvious one discussed before. Nothing is gained by changing frames.

The frame-relative "slowing of a moving clock" is a measured, indirect slowing relative to the frame's global coordinate time. (See Resnick (1968:63, 203); Rindler (1977:28-29, 43-44); Taylor and Wheeler (1992:76-77)). It involves, say, checking the moving clock's "rate" against a series of clocks synchronized according to the frame; it is mediated by the frame's relation of simultaneity. It is not about Jill's clock in itself.

Further, emphasizing frame-relative differences may well confuse. In frame-relative or coordinate time, Jack, in the simple case, measures the *minimum* of all *coordinate* times from P to Q, since he is at rest in the frame in which they occur in the same place. So for any other frame, the coordinate time interval between P and Q is greater. At first glance, the discrepancy runs in the direction opposite to the one needed to explain the different ages. This does confuse people.[10] In Figure 7.1, Jill's journey is approximated by a pair of distinct inertial frames; each of them measures Jack's clock as running slow in the approximating period. That is because each of their global perspectives includes only the early part of Jack's worldline (for one frame) or its late part (for the other). Most of the events that happen to his clock (its "ticks") fall outside the events that get measured in the frames that make up the approximation to Jill's trip.

Any frame's measures of Jill aging are perforce global and not so much wrong as indirect and askew. In 4D language, Jill's cross-sections are measured accurately, but on the wrong angle, from the

---

[10]Taylor and Wheeler (1992:155-56). They explicitly address it as a source of misunderstanding at the foot of 155.

wrong perspective. Although there is nothing subjective in this, there is something relative and it misrepresents her intrinsic time span.

If we stress proper time, proper length and proper mass, then we may rewrite a central message of relativity theory (both special and general) in 3D language and in a metaphysically interesting way: every object has its own intrinsic properties, but there is no one standpoint from which all objects may be directly measured as having these properties. Seen in that light, the theory no longer seems to tell the puzzling message that everything is relative, that there is no special way things are what they are, in themselves. On the contrary. But, nevertheless, the world does not present all this together in any single perspective. What the relativity principles then tell us is which global perspectives make sense, if any do. That is a simple message only in SR: inertial frames make sense.

The main metaphysical upshot of this chapter is that A-theory concepts play no part, nor does a flow of time, in explaining how the twins do it. We neither need nor want the thought that different times must have different ontological standings, the thought on which the flow of time is based. Thus time in spacetime is of the same ontic type as space and spacetime. Temporal relations are relations of separation and connection just as spatial relations are. Despite differences in other respects they unite in spacetime to yield spacetime relations of separation and connection.

8 WHY SPACETIME IS NOT A HIDDEN CAUSE: A REALIST STORY

> Chapter 8 illustrates what it means in GR to reduce gravity to the structure of variably curved spacetime. Crucially, free-fall means a state that is free of dynamical influence. Ideally its trajectories in spacetime are geodesical and this shows just how they are purely kinematical. Since the explanation is reductive it identifies phenomenal and apparently dynamical explanation relative to inertial frames of reference with kinematical explanations in spacetime. Thus the explanatory role of spacetime in GR emerges as without precedent and this establishes that its ontic type is unique.

"Spacetime acts on matter, telling it how to move" (Misner et al., 1973:5); (Taylor and Wheeler, 1992:275).

8.1 Introduction

How does – how could – spacetime act on matter or tell it how to move?

The best short argument against realism runs like this: if spacetime is a real entity for GR then surely the acting and the telling must be a causing – a hidden causing. But, equally surely, spacetime is the wrong kind of thing to make matter move. That's bad physics and bad metaphysics. But if spacetime causes nothing, it explains nothing either. So weed it out of the ontology of GR and settle for a codification whatever that is DiSalle (1994:321-8), (1995:275-7); Brown (2005:24-5); Brown and Pooley (2001:3n). For doubts about codification (in another context) see Nerlich (2005:§2.1, §3.1).

The argument goes astray from the start. Realism doesn't need and can't admit spacetime as causing matter to move. Spacetime is not a hidden cause because not a cause.[1] Yet spacetime *explains* what matter does under "pure gravitation." It does so rather straightforwardly. It exploits various direct identities. That is misunderstood, widely I think, perhaps because the search for causes clouds the issue. That, in turn, may rest on the conviction that thing-to-thing spatial relations can dispense with those of thing-to-space.

Familiar thoughts motivate this chapter. Gravity makes no sense in relativity as action across a distance by some massive things on others. It is not a force, not a cause. GR makes sense only as a local theory: it demands proximal explanation. In pure gravitation situations, the only proximal feature available to explain anything is local spacetime structure. But surely it can't explain matter's motion by causing it. So a style of geometrical explanation, both local and non-causal, merits close consideration. But the idea is alarming. Ontologists abhor spacetime just as nature, it was once supposed, abhors a vacuum.

Apart from its last step, the premises of this motivating argument rest on common ground; indeed, they make up the simplest, basic ideas of GR. The step to the conclusion is no less simple and direct. Further, there are simple examples already to hand of non-causal geometrical explanation.

Handedness depends on whether the containing space is orientable or not. That isn't causal explanation – space does nothing to hands. It is some sort of existential explanation. The shape of spherical space, as a further instance, explains why there are no similar shapes of different sizes in that space – why it has no similarity geometry. For triangles with greater perimeters, there is more space, more area to be enclosed, than in Euclidean space. For spaces of negative curvature there is less. That too, is somehow existential. It's not about causes but about how much space there is and how it is shaped.

What follows in this chapter is familiar and obvious too. So much so that one may puzzle over why it needs to be said. It has not been said. Perhaps the main difficulty lies in the horror of spacetime realism. Dispelling the horror is the hard part of getting it on board in ontology, but that is not closely examined again in this chapter.

I start with some history of inertial motion.

8.2 Cause and classical inertial motion

Confusion once reigned as to what keeps an arrow flying. Galileo's giant stride towards clarity turned on the relativity of motion and the composition of velocities. He saw that "What keeps the arrow flying?" is the wrong question. Instead, ask what causes it ever to stop. Then there are genuine causal answers: e.g. gravity pulls it down to earth,

---

[1] And not hidden either; see §2.5 and ?:38-43.

or it hits something. A more precise message was fogged by the great Italian's preoccupation with circular (including horizontal) motion. This obscured the role of linearity in free motion (Chalmers, 1993).

Newton's first law of motion is clear on linearity:

> Every body persists in its state of rest or of moving uniformly straight ahead, except insofar as it is compelled to change its state by forces impressed.
>
> Newton (1999:416)

Thus Newton straightened out Galileo's story, but only to the extent of Corollary V

> When bodies are enclosed in a given space, their motions in relation to one another are the same whether the space is at rest or whether it is moving uniformly straight forward without circular motion.
>
> Newton (1999:423)

Notoriously, the rest or motion of the "given space" is absolute in *Principia*.

The released bowstring pushes the arrow and causes it to fly. There, cause is force. If we look for a cause why the arrow keeps on flying, we look for a cause of inertial, free-fall motion: we look not for an initiating cause but for a proximal one. That a thing is moving inertially at some velocity *now* might be because it was just moving at that velocity. However, the earlier state doesn't cause or force the later one, despite being a distinct, preceding state. A force is needed to change it. The structure of space is plainly no such cause even though its straights are the specified paths.

But doesn't the preceding inertial motion, the conserved momentum, cause the present inertial motion? Not if we accept both the relativity of inertial motion and that causes are tensors. An adroit frame-swap can transform any state of free motion to a gravitation-free state of rest. The effect vanishes and the cause with it. But a tensor (vector, scalar) that is zero in one frame of reference is zero in all even though its components are non-zero. Cause and effect are tensors and don't vary under change of frame, even when the causal components do.[2] Any further search for a cause of inertial states must look for an account of why things endure. I grant, more for the sake of

---

[2] The first paragraph of Einstein (1905) remarks that, in pre-1905 electromagnetism, causes differ with different uniform motions in unobservable ways. But, as it came to be said, they were mere components of the *same* cause: a 4-tensor. The *components* of the electromagnetic tensor covary under Lorentz transformation. A tensor may be 0 without its components being 0. No cause comes into being with mere change of frame!

getting along than from conviction, that it will be a causal story (e.g. Tooley (1997:398)). Even if there is one, space (spacetime) has no part in it. There is no need to ask why an object at rest in an inertial frame stays put; for any object in uniform motion there is a frame in which it is at rest. The thing merely endures. This is satisfactory: questions can rest at this point.[3]

It's remarkable that the first law says nothing at all about the causal powers of any body to which it applies. It says nothing about what causes anything to endure. The second law requires that all bodies have mass; the first mentions no property whatever. Remarkably, too, we have good classical reason to think that no body ever actually does escape the (gravitational) causal net or persist in its state of rest or uniform motion (although there might be some bodies on which the resultant of forces is zero, briefly or not). This suggests that the law is about trajectories, spatiotemporal entities, not about what might occupy them. It tells us nothing of how any such trajectory ever comes to be occupied. It need say nothing about why an occupant remains on the trajectory, but it does explain why causes are needed to drive it off. It is about the importance for dynamics of the default case: the non-causal trajectory in which there is zero acceleration. The default is rest or uniform motion. I will call these *Galileo trajectories*. Their importance emerges in the relativity of motion and the composition of velocities, which, in turn, depend on the spatial and temporal symmetries of classical mechanics.

The first law really is first. It is conceptually simpler and theoretically deeper than the 2nd. Once we can decide simply when forces are on or off, it identifies the frames of reference (candidate rest states). To a large extent, Newton decided this by seeing free motion as free from impressed forces (impacts, pushes and pulls) and gravity. This laid a groundwork: candidate forces should have (i) observable sources, and (ii) regularities governing (a) when and (b) how they are at work. This rules out arbitrary, conventional postulations of force. Only when we have the right frames of reference and, by implication, the right transformation group, may we explore accelerations relative to them in a comprehensive way; only then can forces be quantified and oriented. Then you can formulate the 2nd law and verify that 2nd derivatives are at the core of dynamics. That the 2nd law entails the first does not rob the first of first place.[4]

---

[3]If all that is sound, then there is a classical non-causal process, a changing of spatial distance between two suitably inertially moving things. We can't cast the motion of either thing as a cause since it vanishes under frame swaps. The rate of change in distance between them is an invariant of the Galilean group: it is a real change. It is identical to the relative motion of the things to each other and not caused by it. It is odd that this never caught on as a clear exception to the rule that all changes are caused.

[4]See Earman and Friedman (1973) for a searching examination of the first law.

8.3 The law of motion

GR's law of motion states that free fall paths are geodesics of spacetime. Trailing clouds of glory from the classical first law of motion, it entered GR as a foundational postulate i.e. the Principle of Equivalence. It reduces gravity to the structure of spacetime; it geometrises gravity. What makes that possible is its uniqueness, regarded as a force. It accelerates all objects in the same way, regardless of their masses and constitution; it has no opposite or negative force and no insulators; it is absurdly weak in comparison with real forces. All this lies behind Einstein's version of Equivalence as it is set in Minkowski spacetime. (see Chapter 9.2.2). It postulates a specific set of paths for all objects that are in free fall. In Einstein's setting it is also obvious, because the "gravitational field" is uniform, that these are geodesics of that spacetime. But this simplistic setting is inadequate for a strong theory of gravity. Whether the geodesical character of free fall trajectories could be retained in constructing the new theory of the non-uniform fields, i.e. the variably curved spacetimes, of GR, was not among Einstein's main concerns in the seven years struggle with general covariance (discussed in §10). Nevertheless, the Principle never lost its place as a leading inspiration of the theory.

Given the complexities of the new theory it was something of a surprise not just that the geodesic principle remained consistent with the fundamental equation but that a formal derivation of it was available. In the view of the Deductive Nomological account of explanation, the law of motion was not only a formal consequence of the field equation but, also, the field equation explained it. But merely that a principle is a consequence of other postulates need shed no new light on its content, especially since Equivalence played so prominent, although informal, a role in guiding the theory's construction.

In Taylor and Wheeler (2000:Chapter 1-7-10; Chapter 3-1-10) the law of motion is derived for both SR and GR from the Principle of Extremal Aging, an action principle.[5] It is essentially a geometric derivation and leads to no advance in our grasp of what it means and why it is true unless we thought the Principle of Extremal Aging more obvious than Equivalence is.

The derivation is not difficult mathematically: for each theory the variational proof has two parts, spatial and temporal. This, too, is a Deductive Nomological explanation of the law of motion but not from the basic equation. The deduction is no doubt revealing about the logical structure of the theory but it hardly leads to a new understanding of the law of motion.

---

[5] The derivation is only for the geodesic law in the setting of Shwarszchild spacetime. There is a general derivation however.

The law of motion is derived from Einstein's main GR equation through its requirement that the covariant divergence of the source tensor vanishes. For details of the mathematical steps see Misner et al. (1973:471-4). The derived equation of motion specifies geodesic trajectories. But it "hurdles" problems of principle along the way (loc. cit). It is a trajectory only for an idealised test particle – one without extension, mass or spin. But no real particle lacks these and other properties. Their worldlines are not identical with these trajectories. The trajectory is geodesical, having everywhere zero acceleration vectors. The complexities of real particles are "hurdled" in the derivation. They are discussed op cit. 473-80 and, in this chapter (Chapter 8.7).

The law of motion is best motivated and understood through the fundamental GR insight of Principle of Equivalence. That develops Galileo's observation that all things fall alike under gravitation irrespective of their masses or constitutions. In GR it morphs into the statement that, given a point in spacetime and a velocity vector there, the worldlines of all point particles so specified would lie on the same trajectory (were there any such particles). That thought is both fundamental and purely kinematic. What then remains to be shown is that such trajectories are geodesics and that is a purely formal matter. Another source of GR's law of motion lies in Minkowski's seeing that the symmetries of spacetime that underlie SR are captured in SR's version of the first law. That is both geodesical and kinematic.

Contrast all this with the derivation of the first law in the context of the "sum over histories" version of quantum electrodynamics.[6] Here it is interference among quantum amplitudes that yields the "law" and only as very highly probable. That indeed puts the matter in an entirely different light. But not in a dynamical one. However, that derivation is not within the ideology of relativity theory and beside the thrust of these arguments.

8.4 Space, time and spacetime in GR

Here's my strategy in a nutshell. In pure gravitation examples, GR explains what matter does by extending the idea of Galileo trajectories to 4-geodesics (straights[7]) in spacetime even though, in general, these have no rest-or-uniform-motion image relative to frames of reference (space and time representations). Roughly that a worldline is a Galileo 4-trajectory explains why its occupant is innocent of causal dynamical dependence beyond its mere endurance (the mere extension of its

---

[6] See Feynman (1985).

[7] I sometimes write "straight" where you might expect "geodesic." Geodesics just are straights of whatever space they are in. The shorter term reminds us of what matters about them for this book.

worldline). It merely falls (floats) freely – free of causes and forces. Of course, the object has causal powers to interact with other bodies and with force fields! However, the images of these Galileo 4-trajectories in the space and time of some reference frame do call up causal and dynamical stories about their occupants since, in that setting, they breach the first law.

This is well illustrated by Einstein's familiar example of a rotating disk in Minkowski spacetime. (Einstein (1919:Ch. XXIII), Einstein (1916:115-7)) Suppose an inertial frame F in which the disk's centre is at rest. Relative to F, a particle in uniform motion crosses the centre of the disk. Its 4-trajectory is Galilean, a straight. Yet, relative to a frame co-moving with the disk,[8] the trajectory reads differently: the particle follows an outward spiral at varying speed. It accelerates. The first law demands the postulation of a force field throughout the frame. It is a gravitational field. It will vanish at the centre of the disk and vary as a function of its radius. Relative to that frame, the particle's motion is forced and caused. Plainly the path and speed of the particle are not uniform relative to the frame. Yet, in the spacetime representation, the 4-trajectory remains straight and Galilean. There is no force on the particle and no cause of its continuing motion. The structure of spacetime explains how the trajectory is Galilean; it does not cause anything.

In GR the "forces" postulated to preserve the first law in the frame may be "real" in the sense that they do not arise simply from a choice of coordinates, but rather from the deviation of geodesics in curved spacetime. These are called tidal forces. They are discussed in some detail in the following examples.

I place two conditions on cause: (i) if x causes y, then $(x \neq y)$; (ii) causes are tensors.[9]

The explanatory role of spacetime in the behaviour of freely falling matter is twofold. It explains (as illustrated) how the apparent gravitational *dynamics* of free-fall particles in general frames of reference vanish into the mere *kinematics* of straights in flat or curved spacetimes. It explains also by citing several identities.

---

[8] In examples like this I use "frame" loosely to mean something like a family of non-intersecting timelike lines with rather flat spacelike hypersurfaces intersecting them all in a spacetime region of no great curvature. Then one gets something a bit like a Lorentz frame and need not worry about how wild the choice of coordinates might be. A local inertial frame is a more strict and quite usual item in the literature.

[9] Following Maudlin (2007) Chapters 1 and 5, I reject the counterfactual account of cause.

8.5 Free fall in a purely gravitational field

Suppose the trajectories of a cloud of non-interacting particles, moving together in a region of flat spacetime, pass into a region of curved empty spacetime. At first the trajectories are both straight and parallel so that the cloud retains its spatial shape. In the curved region the cloud will change shape because the particles' trajectories, while remaining straight, deviate since there is no parallelism in the curved region. Nothing in this implicates spacetime causally.[10] The flatness of the region of spacetime does not cause the curvature of the neighbouring region that the cloud traverses. The curvature does not cause the failure of parallelism: it is that failure. The change in shape of the cloud, the deviation of its point-parts, is the deviation of straight (geodesic) worldlines and not caused by it.

Let's focus more sharply in a frame-relative description of this. Suppose the idealised cloud of matter-points (pressure free dust) is spherical at t_0, falling freely (under gravity alone) towards a massive object. To delete any influence from local matter, assume dust points with negligible mass, ignore gravitational forces between them, and assume there is no other interaction among the points. "Gravity" from the distant source is not erased; it is the curvature of spacetime. The cloud will change shape.

The origin of a space-and-time frame of reference (only locally inertial) floats freely at the centre of the cloud. A point at rest there will remain at rest with zero gravitational force on it. The cloud changes shape round that central point which is at rest in the frame. In the direction of the distant source, the cloud gradually stretches out fore and aft, but it contracts across the orthogonal section – it gets longer and thinner. This is not the product just of choosing misleading coordinates. This is the result of "tidal forces" in the frame: it is how the spacetime curvature projects down into the frame. This closely approximates classical gravity, where it has a causal, dynamical explanation. Clearly, the non-central points move, indeed accelerate relative to the frame. The more distant points acquire larger 3-velocities: some move towards the centre, others away. What accelerates them in the frame is a not entirely conventional force demanded by the 1st law, a "tidal" gravitational force.

A similar tale may be told selecting any point in the cloud as at rest.

That language, that array of theoretical concepts, is appropriate if we conceive of the frame (as we conceive of ourselves) as a spatial thing enduring through time. Spacetime is nowhere in this image.

---

[10]There may be immanent causes for the persistence of the particles: they explain how the cloud gets into and passes through the curved region.

"Spacetime" is not among the concepts in which the space and time explanation may be requested or provided.

But, in a spacetime setting, the changing shape of the cloud is regarded differently. It's explained by the curvature of the spacetime 4-region. The explaining facts must be distinct from what they explain if the explanation is causal. But they are the same facts. In cases of cause, one looks for temporal priority. The curvature of spacetime does not precede the deviation of spacetime straights nor the change in shape of the cloud. True, if the curvature were different the deviation of the straights would be different and so would the shape of the cloud. The conditional is contingent because the identity is. But it is not causal. The priority lies in the direction of reduction and explanation.

Each enduring point is an extended worldline. Spacetime doesn't explain the particle's extension along the straight nor what causes the straight to be occupied (and thus a worldline). Yet Galileo's insight remains – don't seek a dynamical cause for dynamically default states of affairs. That's satisfactory because the straight has a zero acceleration vector at every point. That's what a straight *is*. No acceleration, no force, no cause. It's the default case. That spacetime straights are the worldlines of simple extension is satisfactory for the same old reason – nothing to explain.[11] Here's where question and explanation may halt.

What spacetime explains is only why a 4-straight should be the simplest i.e. the default, dynamical state. It can't explain why anything is in that state.

The identity of the state of affairs differently presented in these descriptions explains what happens to the cloud, so long as there is a cloud. It tells us why the trajectories of the points change the shape of the cloud: the worldlines of different points lie on different straights, and these straights deviate in curved spacetime. The deviating straights project down into accelerating space and time trajectories, among them those that happen to be trajectories of particles. The deviation doesn't cause the acceleration. It's what the acceleration is; it is the change in shape made up by the trajectories of the points. The identities forbid a causal tie.

In turn, the deviation of the worldlines is not caused by the curvature of spacetime, since it is the curvature, in that curvature is the deviation of all geodesics. Flat spaces are those admitting parallels, so "curved space" simply means "space in which straights deviate."

Spacetime doesn't cause material worldlines to lie on straights. If you like, spacetime doesn't fully explain all of this because it doesn't explain the endurance of the test particles. But the endurance doesn't

---

[11] It's not so totally satisfactory that we should assume that we will never find a deeper explanation for it or that the deeper explanation will be consistent with the one made out here.

cause the change in shape. Spacetime explains it through identities, not causes.

This style of explanation through various spacetime identities was without precedent in 1908 (Minkowski) and 1915 (Einstein) and remains unique in science, both physically and metaphysically. Thus it shows the ontic type and role of spacetime as without parallel. That's its metaphysical importance. That makes it a flight of disciplined imagination that is astonishing in its power and creativity.

Finally, to parody Quine – no identities without entities. Only a realist can tell this story.

8.6 Light bending

Eddington confirmed the bending of light rays near the sun as predicted in GR. The immediate observation was of dots on (several) photographic plates of the sun at eclipse. The grouping of dots was caused by a grouping of photons. Our question is about their *separation* and how spacetime structure explains it. We are not concerned with the cause, the source or persistence, of the photons.

To say that light rays bend round the sun is to say that in the 3-space of some frame,[12] light rays do not move along straights of that 3-space. A tidal force, gravity, bends them, in this story.

That's causal. The story is a kind of fiction.

Once more, spacetime is nowhere in this picture. "Spacetime" is not among the concepts in which the space and time explanation may be requested or provided.

In this case, too, when we turn to spacetime, the motion of photons translates up into lightlike (null) straight worldlines. The mapping between space-and-time, and spacetime representations is an identity.

The structure of spacetime explains why the dots on the photographic plates are separated as they are. The explanatory structure is the curvature. Spacetime curvature consists in the deviation of its straights, including lightlike ones. That, in turn, explains the separation of dots on the plate. That's how the photographic surface intersects the deviating luminal 4-straights, independently of whether the plate is there or not. There is no priority in time but only in the order of reduction and explanation. Curvature does not cause the deviation of all straights because it is that deviation. The curvature tensor simply analyses and measures the deviation – an identity not a cause. Flat spacetime is unique in having parallels. The failure of parallels is the curvature: it is the deviation of straights.

---

[12]Not an inertial frame, since spacetime is curved and lacks parallels. Only in the limit is spacetime flat and inertial frames locally available.

That completes the explanation. It is not causal; it is realistic –
no identities without entities.

8.7 About matter

I've told my story with some idealised bit-players – test particles. My
cloud of dust was misrepresented as made of massless particles each
of which tracks a straight in a structure unaffected by these contents.
But real dust is made of small but extended specks, not particles.
Even specks have some mass that will constrain spacetime structure;
clothed with specks, spacetime doesn't have the same straights as it
has naked.

Any spacelike cross-section of a speck worldline will be intersected
by more than one timelike straight. Since these deviate in curved
spacetime, the causal story within any speck is not trivial. Elastic
forces inside resist the deviation of the speck's smaller parts: internal
stresses, distortions, will arise in it. As elastic wholes, specks' worldlines won't be straight trajectories although there might chance to be
a straight that lies everywhere inside the speck.

The causal story about specks is exhausted in the play of electromagnetic forces engaged in resisting the distortions and in any immanent causes of speck endurance. As before, spacetime explains the
deviation of geodesics that change the electromagnetic forces, but it
does not cause the forces. It continues to explain as before the causal-
default part of the story – why this trajectory needs no cause for any
geometrically simple, purely time-extended thing to lie along it. At
each point, its space-like acceleration vector is zero. The spacetime
story is about the cause-free status of the trajectory. There is no dynamics in the tale. That explanation does not encroach on any theory
of matter. An occupying point is irrelevant save as an illustrative
fiction.

Yet we do accept exactly that explanation in real if approximate
cases. The orbit of Mercury is calculated treating the planet as a
point (among other approximations). The observed advance of the
planet's perihelion, famously, is very close to the GR-predicted Galileo-
trajectory along which the idealised planet would extend. The orbit is
a spacetime straight. Unknown stresses within the planet, and unob-
served imperfections in its straight 4-trajectory are ignored. We fully
understand why the orbit is the one we see: it's virtually a geodesic.
It exhibits, it traces, the structure of spacetime. The structure is not a
hidden something (not concealed, not obscured, not too small, not too
fine). It is observed with highly non-trivial precision, even though we
know that we see an approximation and that the unoccupied straight
itself is not a visual object.

Again there is an order of reduction and explanation but not of temporal priority.

Spacetime, therefore, is neither a cause nor hidden. We can understand matter as tracing the structure out. Einstein was concerned about the dependence of the metric of spacetime on rods and clocks, mechanisms that are essentially alien intrusions into the theory. In a notable paper (Ehlers et al., 1972) it was shown how the evidence for the metric, its manifestation one might say, can be found in light and particle interactions. Briefly, the light cone displays the conformal structure of spacetime (i.e. the null cone structure), the trajectories of free particles display its projective structure and their interactions yield the affine structure and, finally the metric. What needs to be said at this point is that this tracing of structure, like Mercury's tracing out of its spacetime geodesic, is an approximate idealisation. As before, test particles are a fiction: real particles only approximate the projective structure in their motions and so on. The pencils we use are rather blunt. Nevertheless, the claim of these interactions to lay open to view the geometric structure of spacetime can hardly be damned as an extravagant fancy.

For similar reasons, in illustrative explanations, we may ignore the epicycle of feeding the small masses of the specks back into the **T** tensor. That will simply generate a new set of straights and these will be causal-default trajectories as before; the geometric explanation exploits just the same feature of the revised spacetime structure and its straights. It wasn't really ever about the properties of matter.[13]

8.8 A parody of "hidden cause"

Einstein regarded GR as correcting crucial flaws in the role of the principle of inertia in classical and SR physics. In both, the principle "seems to compel us to ascribe physically objective properties to the space-time continuum ... [but] it is contrary to the mode of thinking in science to conceive of a thing ... which acts itself but cannot be acted upon..." (Einstein, 1953:55-6). He believed that GR yields the desired result that spacetime's action on matter is matched by matter's reaction on spacetime. There is a kind of mantra throughout Misner et al. (1973) and Taylor and Wheeler (1992): spacetime acts on matter telling it how to move: matter acts on spacetime telling it how to curve. The arguments of this chapter entail that there is neither action nor reaction between the two. The mistake is crucial both here and in the

---

[13]Compare (Brown, 2005:24) that "... world-lines [of test "particles"] follow geodesics *approximately* and then *for quite different reasons*" from anything to do with the nature of test particles (his italics). Apart, of course from their natural tendency to persist. That leaves the story told here untouched.

arguments of the last chapters.

Brown and Pooley begin their (2006) with a discussion of the principle of action and reaction. It leads them to a critique of spacetime as playing a dominant explanatory role in Minkowski's geometrical theory. They assume the terms set out by Einstein – there would have to be an action of spacetime on matter. This casts geometrical explanation as dynamical explanation. In a spirit of irony, they invent a preposterous theory, according to which spacetime geodesics are likened to gutters or grooves along which spacetime "nudges" things. They ascribe this theory to Einstein and, in search of another peg to hang it on, to me. In the latter case, they cite a brief passage about something quite different – that explanation in GR is local.[14] The tacit premise of their preposterous theory is that spacetime explanation simply must be somehow dynamical. Since only a daft dynamics could do so strange a job, it is pressed into service as a man of straw. Ironically, the core message throughout Nerlich (1976, 1994) is that spacetime does not act: geometrical explanation is both needed and available but it is neither dynamical nor causal. Spacetime could not and does not act on matter. Brown and Pooley merely deride the attempt to make geometrical explanation intelligible and plausible.

In a geometrico-causal story, they argue, matter must follow something like "grooves" or "gutters" in spacetime along which spacetime "nudges" them (Brown (2005:24, and 161) for "nudge"). The thought is that the grooves force things to follow them. This geometric story is indeed both causal and absurd.

Despite their calling this view popular, there is no published version of anything like it, although it sometimes – too often – comes up in discussion. It is quite unworkable; how could it yield the crucial result that the geodesic followed in free fall is independent of the mass of the falling body? But something makes this mistake easy; it is exactly what makes the argument I mentioned at the beginning of the chapter so plausible. Doesn't all explanation just have to be some kind of causal, dynamical explanation - perhaps because spatial relations cannot exist without bodies? So geometric explanation must conform.

Two interesting points arise: (i) the parody presupposes that test bodies would be doing something else if the nudge along spacetime's grooves did not turn them from it. Without that presupposition, the gutters, the nudges and the parody itself gain no intelligible traction. (ii) this never-mentioned something else would either be a state without external cause or have one. If it is uncaused, some causal default state is tacitly admitted as intelligible: why not the state we began

---

[14]The passage, Nerlich (1976, 1994:264) says nothing about geometric or kinematical explanation and nothing about gutters or grooves, but something about explanation's being local in GR. Brown and Pooley have conceded the misinterpretation in private communications. Nothing is in print to this effect.

with? If it is an externally (e.g. electromagnetically) caused state, then GR tells us that the trajectories won't be geodesical after all, and the grooves would play a totally obscure part in the theory. I conclude that the theory *presupposes* the intelligibility of causal-default states, so that Brown and Pooley presuppose it, too. I agree, of course. Nothing suggests that we should know the states *a priori*. My colleague, Greg O'Hair suggested that the default might have been a random spatial walk. It is an empirical, theoretical fact that the causal default is a spacetime straight. The identities cited before are also empirical and theoretic. That makes perfect sense within a contingent geometry and mechanics. I claim that it is satisfactory. It is not dynamical nor causal in that sense.

Brown's own view of this stresses the fact that, in GR, the law of motion is entailed by the field equations. This is a significant and unprecedented formal aspect of the theory. But what is its explanatory significance? Brown speaks of free particles' trajectories as explained by an entirely different reason from the one I have given. He objects "it cannot simply be in the nature of free test particles to "read" the projective geometry or affine connection or metric, since in the general theory they follow geodesics only *approximately*, and then *for quite different reasons*" (Brown, 2005:24).

Indeed, it is not in the nature of particles to read structures or anything else, for free particles do nothing. We also saw that real specks of dust are not free from inner stresses that may drive them off the ideal trajectory. But Brown means something more. He does not say directly what it is, referring us, instead, to Misner et al. (1973:Chapter 20.6). This takes us to a discussion of the derivation of the law of motion but, more directly relevant, to a discussion of the behaviour of real particles, including their quantum properties, in a gravitational field. The message there, put informally, is that when the properties of particles with mass, angular momentum, quadrupole moments and so on are fed back into the matter tensor, curvatures highly local to the particle develop a very complex geometry so that the geodesics not merely deviate, but do so locally in complex ways (Op. cit. 473-480). This means that the dynamics within the particle, the inner stresses set up by the deviations, as more generally described in the last paragraph of the last section above, will play a role. These forces would nudge the particle, not along the geodesic, so to speak, but off it. But, in principle, the situation is of the same kind as in the case of Mercury. The explanatory role of spacetime is still to mark out geodesics as dynamically null trajectories. It explains by describing kinematics.

For Brown, the "mystery of mysteries" ((Brown, 2005:Chapter 8.3.2); see also pp. 14-16, 23-6, 140-42) is how spacetime can act on free particles or how free particles could conspire to follow geodesical "grooves." But since spacetime doesn't act and particles don't conspire, we need not ask how or why. But we can usefully ask a slightly different question: what dynamical state must something be in for its worldline to be a geodesic? Geodesics are spacetime trajectories on which, at each point, acceleration vectors are zero.[15] So a thing whose worldline is straight must either have no accelerating forces acting upon it or have the resultant of forces zero. Otherwise it will have an acceleration vector and deviate from being straight. That describes a dynamical null case; i.e. a purely kinematical one free from dynamics. Further, the geodesical case is never an absolute motion: the point can always be taken as at rest in a local *inertial* frame. Force-free objects don't run along spacetime grooves or follow straights. They do nothing and nothing is done to them. They just stay put, unless they cease to endure. In SR, spacetime structure allows straightforward cases of force-free motion of extended bodies, since the worldlines of all of its points can be parallel straights. But in GR nearby straights usually deviate: This makes things more complex. Thus one way in which spacetime constrains the behaviour of bodies is through the geometry of its vector fields.

It's true that the structure of spacetime is contingent and its identity with geodesical trajectories is a contingent identity. It then seems obvious that the structure contingently "makes a difference" to what happens. I mistrust the clouds of meaning trailed by "makes." It is a difference but there is no making.

8.9 Does matter act on spacetime, telling it how to curve?[16]

There are two bits of unfinished business. Identity arguments are powerless to settle two remaining problems. (i) They can't relate geometric structure in one spacetime region to that in a distinct region nearby. But I do not think that is a causal relation either; (ii) If spacetime can't act causally on matter then matter can't act causally, dynamically, on spacetime. Identity arguments look impotent to tell us how matter is related to spacetime in that direction. The field equation is not an identity. Nevertheless, it is not causal either, but an equation of mutual formal constraint. One aspect of the identity of gravitational with inertial mass is that GR need only consider inertial mass. The left hand side of the field equation need not be taken as a source term. This needs long and careful reflection on the relation between curvature of spacetime, gravitational energy and the mass of curved, empty spacetime and more. That task is taken up in Chapter 9. Schrödinger

---

[15] In the present 4-dimensional context, an acceleration vector means a vector directed away from a timelike geodesic, not one that increases frame-relative speed.

[16] Misner et al. (1973) say yes.

suggested that the mutual constraints placed by each side of the basic field equation on the other should be understood not as an action of the one on the other but as an identity:

> I would rather you did not regard these equations as field equations, but as a definition of T_{ik}, the matter tensor. Just in the same way as Laplace's equation div$E = \rho$ ... says nothing but: wherever the divergence of E is *not* zero we say there is a charge and call div$E$ the density of charge. Charge does not cause the electric vector to have a non-vanishing divergence, it *is* this non-vanishing divergence. In the same way, matter does not *cause* the geometrical quantity which forms the first member of the above equation to be different from zero. It *is* this non-vanishing tensor, it is described *by it*.
>
> Schrödinger (1950:99) Original emphasis.

The thought is both intriguing and alluring, but I find it more congenial to note that GR is not tied to any particular theory of matter. Schrödinger's suggestion might be read as envisaging **T** as a placeholder for some suitable theory of matter where the constraint of **T** on **G** is an issue awaiting clarification.

I take the geometric style of explanation of the motion of matter set out in this chapter to provide a criterion for whether a field theory of matter has been unified or geometrised. If a theory's law of motion selects geodesics as its paths so that what appear to be dynamical features of that field theory are displaced by kinematical features and cause is displaced by appropriate identities as already argued then it is geometrised. The gravitational field satisfies this condition and, so far, it is the only field theory that does.

That leaves the problem how spacetime "acts on matter" unsolved. I do not know how to dig deeper into this patch of ground.

9 Is Spacetime Really Spacetime?

Chapter 9 begins an exploration of theories that take spacetime to be other than it seems. Einstein began this tradition. It also has some modern adherents. Various suggestions are explored: that it is an ether of a novel kind; that the metric field of spacetime is really a kind of force field not to be distinguished clearly from dynamical or matter fields. But while the metric field really is a field it is utterly unlike any other field, as explained in chapter 8. The idea of gravitational energy is explored and found wanting. These attempts to rebottle the genie spacetime all fail. The conception of spacetime as a concrete immaterial thing is maintained.

9.1 Introduction

Einstein's first GR paper (Einstein, 1916) has two distinct parts. §§1–3 introduce the theory in broad terms and make a number of philosophically interesting claims. From §4 on, the mathematical physics (i.e. the theory proper) is developed in now familiar detail. It is this second part that delivers the reduction of gravity to spacetime geometry, especially to its curvature, and firmly commits GR to the existence of spacetime. The opening sections reveal that Einstein thought rather differently of his achievement than the ensuing, path-breaking, formal development can justify.

From its first appearance, spacetime was an uneasy presence in GR. The first 3 sections of this great paper give a rather philosophical overview of it as a *relativity* theory, satisfying to relationists and plainly congenial to Einstein. His most striking claim is that the new coordinate style of general covariance (the topic of the next chapter) "takes away from space and time the last remnant of physical objectivity" (op. cit. §3; 117). That surely denies spacetime a role in the theory.

A major error about general covariance was quickly pointed out in Kretschmann (1917) and acknowledged by Einstein (1918): all earlier theories of mechanics, whatever their base space, can be formulated in generally covariant style. That style concerns just the range of permitted coordinates. It has little to do with the structure of space, time and spacetime. I conjecture that this error convinced Einstein that general covariance did not take the last remnant of physical objectivity from space and time. In that case §4 to §22 of (1916), left as he had written them, really did reduce gravity to the structure of spacetime and thus committed GR to its existence. This would have been a most unwelcome conviction. I guess that it led him to recast GR in order shed that commitment.

Four years later, in the Leyden lecture in 1920 (Einstein, 1983:Chapter 1) he argued that spacetime is really a kind of ether. In the long Appendix to the fifteenth edition of his popular presentation (Einstein, 1954), he argued that spacetime's metric field is really the gravitational field, thus reversing the widely accepted reduction of the gravitational field to the structure of spacetime. The second part of his (1916) carried that out brilliantly and no fault in its maths or physics has damaged it. It was just this reduction that was so striking, bold, elegant and fruitful. It made the theory famous.[1] Despite this a number of distinguished modern physicists and philosophers of science have expressed reservations similar to Einstein's, not always for the same reasons.

Underlying all these worries about spacetime is this: its explanation of what happens in free fall requires that gravity be reduced to spacetime as a concrete, insubstantial particular entity. That requirement has never come into sharp focus so no attempt has been made to explore in depth what that description could mean or how it could be correct. The result has been a widespread, continuing hostility to spacetime. My aim is to complete that exploration.

The first part of this chapter examines Einstein's arguments both in the Leyden lecture and in his popular exposition. (ops. cit.). It also considers related but more recent arguments.

The word "physical" clouds all this debate. Despite its frequent occurrence in the debate, it lacks a sharp-edged meaning in the literature. I call spacetime physical because it plays a key role in GR, a notable theory of physics. GR is a reductive theory in which spacetime structure explains all that gravity once explained and more, transforming the dynamics of the latter into the kinematics of the former. Chapter 8 explains how: spacetime clearly remains itself and plays a kinematical role, not a dynamical one; spacetime has no causal powers and is

---

[1] That has long been my opinion and it was pleasing to discover that it was Wheeler's opinion too. See Wheeler (1998:230-1).

not a substance in that key sense. Granted that, then one physical entity whose properties figure prominently in explanation in GR really does differ in ontic type from the standard concepts of dynamics. To say that spacetime is a physical entity, especially in this context, may insinuate, intentionally or not, that its ontic type is not unique in physics. That would mean that a seamless array of concepts go to make up the content of GR and other theories of mechanics. I always mean by "physical" only that spacetime is an item in GR and other theories of mechanics. It is sometimes far from clear whether this is all that others mean by it.

A crucial element in contention here is Einstein's enthusiastic welcoming of GR as satisfying a principle something like Newton's Third Law: there is an action of spacetime on matter and a reaction of matter on spacetime (Einstein, 1953:55-6). But spacetime does not act on matter and it is uncertain how matter constrains spacetime. The former is well explained in the second half of Einstein (1916). The latter remains poorly understood.

9.2 The ether: Einstein and others

Einstein's views on aspects of GR varied little over the years. His rejection of spacetime as a real concrete immaterial non-dynamical thing was rather constant. In this section I consider two statements of his views, the first given in a lecture in 1920. The second was written late in his life, in 1952 (Einstein, 1954:Appendix V). Neither discussion is very technical. They are rather general reflections without complex detail from physics or mathematics.

In a preface to his (1954) he states the aim of the concluding Appendix as follows:

> I wished to show that space-time is not something to which one can ascribe a separate existence, independent of the actual objects of physical reality. ... In this way the concept "empty space" loses its meaning.

Einstein (1954:vi).

9.2.1 The concept of field

The argument begins with the concept of a field and, in particular, with what Einstein called the pure gravitational field (op. cit.: 154). It is best not to nail oneself to a single formal definition of "field" since a neutral view is needed to avoid at the outset either restricting or evacuating what Einstein had to say. I turn to an authoritative introduction to fields that begins with the familiar idea of a fluid: "A fluid is represented as a continuum characterized by physical quantities that may vary smoothly from point to point and from moment to moment" (Torretti, 1999:168). Then a field itself is some physical condition, its quantities smoothly varying from point to point throughout space and time. Some presuppositions of both this and Einstein's summary are contentious: what is meant by "physical"? What is conveyed by the phrase "throughout space and time" – is the condition or property spread on spacetime as paint is spread on a blank canvas or is the condition or property an attribute of space or spacetime (and its points)? A look at Einstein's argument will make clear why these are vexed questions.

Fields took on a fundamental importance with the advent of electromagnetism. A compass needle pointing along a wire turns at an angle to it when a current is passed. That was the first sign that one might need to think of electricity as an extended condition not only in the wire but also in space surrounding it, a condition that came and went as a current flowed or ceased. The idea that non-contact forces such as gravity could act between bodies only in a centre-to-centre direction was dead and the rich field concept thereby born, alive and kicking. Electromagnetism is still the paradigm case of a field.

Maxwell's crucial further development of the field concept emancipated it "from the assumption of its association with a mechanical carrier." Einstein describes it as among "the most psychologically interesting events in the development of physical thought" (Einstein, 1954:146).

9.2.2 Principle of Equivalence

The crux of Einstein's argument comes with his account of the Principle of Equivalence. Since not all readers will be familiar with it, here it is in full:

> We start from an inertial system S_1, whose space is, from the physical point of view, empty. In other words, there exist in the part of space contemplated neither matter (in the usual sense) nor a field (in the sense of the special theory of relativity). With reference to S_1, let there be a second system of reference S_2 in uniform acceleration. Then S_2 is thus not an inertial system. With respect to S_2 every test mass would move with an acceleration, which is independent of its physical and chemical nature. Relative to S_2, therefore, there exists a state which, at least to a first approximation, cannot be distinguished from a gravitational field. The following concept is thus compatible with the observable facts: S_2 is also equivalent to an

"inertial system"; but with respect to S_2 a (homogeneous) gravitational field is present (about the origin of which one does not worry in this connection). Thus, when the gravitational field is included in the framework of the consideration, the inertial system loses its objective significance, assuming that this "principle of equivalence" can be extended to any relative motion whatsoever of the systems of reference. If it is possible to base a consistent theory on these fundamental ideas, it will satisfy of itself the fact of the equality of gravitational and inertial mass, which is strongly confirmed empirically.

Op. cit. 151-2.

Here no structure is added to the objective, absolute 4-dimensional world by changing from an inertial reference frame to an accelerated one and then speaking of a gravitational field. Throughout, Einstein's argument is set in a simple, flat Minkowski spacetime. We need no gravitational modification of that *spacetime* geometry to cope with the mere change of frame. True, we must talk of gravity when we turn to non-inertial frames of reference for our description of what happens. In such frames the first law can't be maintained unless gravity "forces" the motions that are non-uniform relative to the frame. But in a spacetime perspective, nothing has happened – the introduction of the frame S_2 has obliged us to re-describe, trivially, test particles that are at rest or uniform motion relative to S_1, as accelerated in the new frame. The crucial feature of gravity is that all objects are "accelerated" alike regardless of mass or constitution. That obligation is imposed directly by the bland geometric structures of space, time and spacetime. It involves no more than that. There is a gravitational field relative to S_2 as a mere artefact of the frame geometry. But spacetime remains as it was.

This was probably Einstein's first step towards a view of gravitation as a field, a field identical to the metric field of spacetime. It is remarkable that he regards inertial systems as losing objective significance whereas the argument strongly implies that gravity loses it. The "field" described has no source and is, impossibly, *uniform*. Its before-mentioned peculiarities as a force and its complete absence from spacetime structure are what allow us to regard its frame representation as a coordinate artifice. While that is an entirely new picture of what gravity might be, no new matter physics has been introduced. It remains a purely geometric picture set in Minkowski spacetime. There, inertial frames of reference do remain privileged until a full theory of gravity can be introduced. There is no significant gravity consistent with SR.

Nevertheless, this was a highly significant step. Since all Riemannian spacetimes are locally flat, there are locally usable inertial frames in curved spacetimes but no global ones. The local ones are only approximations.

9.2.3 Action at a distance and the ether

Einstein, in his (1983) takes us to a view of spacetime as an ether. Of Newton's action-at-a-distance theory of gravity he writes (p. 4-5)

> ... this theory evoked a lively sense of discomfort among Newton's contemporaries, because it seemed in conflict with the principle springing from the rest of experience, that there can be reciprocal action only through contact ... How was unity to be preserved in his comprehension of the forces of nature? Either by trying to look upon contact forces as being themselves distant forces...; or by assuming that the Newtonian action at a distance is only *apparently* immediate action at a distance, but is in truth conveyed by a medium permeating space, whether by movements or by elastic deformations of this medium.

This second disjunct got nowhere at all in Newton's day but, clearly, Einstein wished to revive it. And, in GR, geometrical explanation is indeed local but involves no forces. The Leyden lecture opens as follows:

> How does it come about that alongside the idea of ponderable matter, which is derived by abstraction from everyday life, the physicists set the idea of *another kind of matter, the ether*? The explanation is probably to be sought in those phenomena which have given rise to the theory of action at a distance, and in the properties of light which have led to the undulatory theory.

> Op. cit.: 3: my italics.

SR is not inconsistent with every picture of ether (though it does conflict with the classical Lorentz ether); in his (1905), Einstein claimed only that the ether was superfluous. But, in GR, it wears a different aspect. The metric can't, in general, be written in the familiar Lorentzian form because the geometry is wrong for it. It must be written with coefficients $g_{\mu\nu}$ as factors of the coordinate differentials $d_{\mu\nu}$. First, this cancels the familiar connection of coordinate differences with direct measurements using rods and clocks (op. cit: 154-5). Second it is formally constrained by the distribution of mass-energy, the tensor field $T_{\mu\nu}$.

What is fundamentally new in the ether of ... [GR] ... as opposed to the ether of Lorentz consists in this, that the state of the former is at every place determined by connections with the matter and the state of the ether in neighbouring places whereas the state of the Lorentzian ether is conditioned by nothing outside itself, and is everywhere the same. The ether of ... [GR] ... is transmuted conceptually into the ether of Lorentz if we substitute constants for the functions of space which describe the former, disregarding the causes which condition its state. Thus we may also say ... that the ether ... [GR] ... is the outcome of the Lorentzian ether through relativation [sic].

Einstein (1983:19-20).

But this does not explain just what role an ether might play. Einstein wished to provide for "... a unity to be preserved in [a] comprehension of the forces of nature" (loc. cit. above). He wanted *reciprocal action only through contact* as above. The ether emerges as having no role but to mediate an action of spacetime on matter and vice versa. It would permit action and reaction to be local and unify forces. But, in GR as a force-reductive theory, neither forces nor the third law of motion have purchase: there can be no *reciprocal* action encoded in the fundamental equation because there is no *action* in the first place. There is a non-action, no force, explanation that makes local sense as kinematic. (Misner et al., 1973:5) and (Wheeler, 1990:Chapters 5, 6) rightly take spacetime itself to provide local explanation in GR yet they still allow spacetime to "grip matter and tell it how to move." That is ontological obscurity as was shown in Chapter 8 where we saw how spacetime's structure provides the desired style of explanation. If that succeeds, the ether, once more, is a superfluous unobservable obscurity. Further, spacetime is indeed there but empty if, everywhere and when, $\mathbf{T} = 0$.

Later, Einstein (1954) argued that spacetime "... is not something to which one can ascribe a separate existence, independent of the actual objects of physical reality." He draws a contrast between space and anything that is a field – "that which fills up space and is dependent on the coordinates" is a content of space (op. cit. 155). Thus the gravitational field (ether?) is an actual object, a field filling up space with the $g_{\mu\nu}$ as its field potentials and depending on the coordinates. But then to remove that field is to remove everything can be regarded as spatiotemporal. It removes Minkowski spacetime since, even in that trivial case, each $g_{\mu\nu} = 1$ or 0 (in Lorentz coordinates). So each $g_{\mu\nu}$ is cast as a field potential, not a structure of spacetime. But to regard the $g_{\mu\nu}$ as potentials of the gravitational field as well as coefficients of coordinates in the metric field plainly *inverts the original reductive intention to reduce gravitation to the structure of spacetime* without giving any clear reason for doing so.

He claims (1954) that this also removes topological structure. It does not. There is an indispensible topological base space for GR, the differentiable manifold. It belongs in the same troublesome ontic type as space and spacetime. The genie eludes this attempt to rebottle it. That is discussed at length in the next chapter.

Einstein's arguments for an ether and for spacetime's existence as dependent on matter fail. They invert the explanatory thrust of GR. They abandon the conceptual daring, the deep insight that belongs to its first formulation. They misrepresent the style of explanation that the Principle of Equivalence requires and delivers. The upshot is a more complex, cumbersome (saddled with the obscure ether) and more traditional theory – a disappointment. It cannot be overemphasised that spacetime *does not act on matter* nor vice versa. The crucial underlying thought about the relation between matter and spacetime, that an action requires a reaction, has no application in GR's picture of free fall and gravitation.

9.2.4 Two modern views

John Earman (1989) and Robert Rynasiewicz (1996) also argue for taking the ether with a touch of seriousness. In each, the case is made for a dematerialized ether. I confess that I am not clear just what is intended by that. In the case of Rynasiewicz the aim is clear enough – neither the concept of spacetime nor that of a dematerialized ether can be sufficiently sharpened so as to present us with a clear decision whether GR is committed to one or to the other. Rynasiewicz's paper includes a scholarly survey of the various guises the ether has assumed in its career in physics and he concludes that it remains obscure how to distinguish it from spacetime. Thus the traditional debate between relationism and substantivalism is outmoded. As I understand this, the vagueness lies mainly on the side of the ether and the argument does not bear clearly on the spacetime realism espoused in this book. (See Dainton (2010:379-381) for an outline of this and, for a critique of it, Hoefer (1998).)

Earman sees the differentiable manifold as the ether – that is, as the basic bearer of properties in GR. So the ether is a spacetime object, yet not the full-blown metrical thing we usually understand spacetime to be. That conclusion would follow from the hole argument. It is examined in the next chapter.

But Earman also sketches a different case for a friendly view of the dematerialized ether appealing to what he calls the holing operation (Earman, 1989:Chapter 8.4). This is rather like the surgery suggested by Bricker and discussed in Chapter 2 §4. If spacetime is real, then it can have surgical holes cut in it, points removed and so on. I take the suggestion to be that spacetime is a substance in some sense richer than a mere subject of geometric attributes. Perhaps it is the view that a real spacetime has causal powers. But nothing in realist metaphysics obliges it to commit to this or to properties once thought necessary features of a space. It need not be Euclidean. It may be twisted, curved and, more to the present point, it may have bounding surfaces or edges, cusps, "missing" points and holes. Nothing about space as modestly conceived by the arguments of Chapter 2 §7 forbids a space or spacetime with surgical holes. (Presumably, no one thinks there is an *operation* of cutting such holes.) Earman's discussion of the point is geometrical throughout. Witness the definition of a surgical hole that Earman offers (op. cit.: 160) to substantivalists. It is congenial to the account of realism given in Chapter 2 §7.

By contrast, the metaphysics of substantivalism as accurately reported in Earman's definition of it (op. cit.: 11) is an obscure doctrine.

A main point at issue is the style of explanation in GR as characterised in Chapter 8. It identifies a clear boundary in GR as to which kinds of explanation are dynamical and which are purely kinematical. As it has often been put, construing gravity as a force, i.e; as dynamical, overlooks the features of it that are alien to forces. Free fall 4-trajectories are all and only those with zero acceleration vectors. Were any particle able to have such a worldline, no dynamical properties explain why the 4-trajectory is geodesical. If we take spacetime straightforwardly as having determinate geometric structure up to an affine connection, the whole explanation can be framed without recourse to dynamical properties. We can get purely kinematic explanation up to the affine behaviour of free fall objects. *That is, gravity can be and has been geometrized.* While we can retreat from these bold steps by re-introducing "gravitational forces," the point where we must turn to dynamics is sharp. The key question is: which fields can be geometrized? The short answer is: none but the metric field.

On the face of it, prediction and explanation in electromagnetism require force functions as well as coordinate ones. That is not quite straightforward however since attempts to geometrise the electromagnetic field have had modest success. Rainich was able to geometrize the charge-free electromagnetic field and Kaluza-Klein did so by resorting to a fifth dimension (for details see Graves (1971:Chapter 15)). The relevant issue here is not whether these attempts succeed or fail but whether there are well-defined criteria for success in geometrizing fields. These are the criteria by which Einstein failed, for so long, to find a unified field theory. These criteria also provide for defining a maximal role for spacetime structure *in explanation in standard GR*. That is sufficient for my purpose. There is no point in invoking a nebulous concept such as dematerialized ether.

Perhaps the distinction is doubted because, formally, the role and structure of fields look alike in GR. But only in the "gravitational field" do all objects fall alike regardless of their masses and what they are made of. Every massive entity has a "gravitational field" and has it under all conditions. Run a comb through your hair. Then, but not before, it will pick up light pieces of paper; you can bang an ordinary magnet with a hammer and weaken its power to attract and repel. You can't knock the gravitational attraction out of anything. Finally, considered as a force, gravity is absurdly weak. These comments hold roughly but significantly. The metric or "gravitational field" is formally, but only formally, like the forces in the standard model. Arguably (Petkov, 2012:Appendix C) physicists should cease to hanker after a quantised gravity. There is no gravitational force or gravitational energy.

9.3 The fundamental equation

The first two subsections of this section are largely an expository prelude to the discussion of gravitational energy. Some readers will want to skip it.

The field equation of GR reconstructs the classical relation between the gravitational potential field and mass-energy density, ρ. The new equation is most briefly expressed as:

$$\mathbf{G} = k\mathbf{T}$$

or, in coordinates, as :

$$G_{\mu\nu} = kT_{\mu\nu}$$

ρ is not frame invariant so a new invariant object, $\mathbf{T}$, had to be constructed. $\mathbf{G}$ is designed to equate to the newly designed $\mathbf{T}$.

The result is an equation not an identity. Read straightforwardly, it takes for granted an exclusive distinction between geometry and matter, or between spacetime and matter fields. If so, the distinction is there from the beginning and the structure of GR builds upon it. It is an equation of formal mutual constraint. If $\mathbf{T} = 0$, then $\mathbf{G} = 0$, but while the former equation states the absence of matter, the latter by no means states an absence of spacetime structure, admitting a wide variety of them. $\mathbf{G}$'s being 0 somewhere does not entail that the curvature tensor, Riemann, is 0 there nor that spacetime is Minkowskian or nicely flat anywhere. The Weyl rank-4 tensor, roughly a complement of Ricci, describes the structure of spacetime where $\mathbf{T} = 0$ and there are no matter fields. The absence of matter does not entail an absence of gravitational physics or "gravitational energy" – so called.

A sketch of the content of the two sides of this equation follows.

9.3.1 The Einstein tensor G

The tensor that measures curvature is the rank-4 Riemann, $R^\alpha{}_{\beta\gamma\lambda}$. Since **T** is a rank-2 tensor (a contraction of Riemann) some rank-2 tensor, **G**, must be found for the geometrical side of a formally possible equation. Ricci is a direct contraction[2] of Riemann, an averaging of it over all directions. So **G** will include Ricci as its main component. Many tensors involving Ricci meet the formal requirement; but a further condition crucial for the local conservation of the source, has terms involving the metric. (For more detail see Misner et al. (1973:40-43; Chapter 8.7), Schutz (1985:174, 196-7), Wald (1984:Chapter 3.2).) This leads to defining **G** as a symmetric tensor, the sum of Ricci and a product of the metric with an undetermined constant, λ. It is:

$$G_{\mu\nu} = R_{\mu\nu} + \lambda g_{\mu\nu} = G_{\nu\mu}$$

It is so constructed that its covariant divergence vanishes automatically i.e. $G_{\mu\nu;\nu} = 0$. This vanishing is enforced on both sides of the equation, expressing a conservation principle.

9.3.2 The matter tensor $T_{\mu\nu}$

A sketch of the matter or stress-energy tensor follows.[3]

A large range of problems in GR may be neatly handled in terms of the dynamics of relativistic fluids. These permit interesting but simple examples of $T_{\mu\nu}$. A fluid is one remove, so to speak, from a continuum. It is composed of particles so numerous that it is impractical to trace them singly and one must consider averages – particle number, mass and momentum densities – in volumes small enough to be regarded as homogeneous i.e. as having a useful density, average velocity etc. The "streaming" of 4-momentum in spacetime can thus be captured. An *element* of the fluid may be visualised as a small spatial box. **T** describes its contents, and changes of contents over time. For the purposes of differentiation, elements may be regarded as points, despite describing their components as small volumes with surfaces. A perfect fluid is one in which the only interactions between elements are push and pull forces. Slide-resisting forces (viscosities) are zero. There is no heat conduction. The simplest fluid is dust, a cloud or collection of test particles in which each is at rest in some inertial frame, F. The number of particles per element may vary. If the element is a unit volume, the element's *number density, n,* in its rest frame is just the number of particles in it. Its *mass density* is simply the product of the number density and the average mass per particle, mn. In the

---

[2]In a contraction of Riemann, the first upper and and last lower indices are identified and the result summed to yield Ricci.

[3]I follow Schutz (1985:§4), but see also Misner et al. (1973:§§5.1 − 5.7).

rest frame, the energy density is just the mass density and there is no momentum. That gives an analogue of the classical mass density, ρ, the source of the gravitational field in Newtonian gravity. ρ is not an invariant but merely the T^{00} component of **T**.

In a frame, F, relative to which an element moves, its volume is Lorentz contracted, so the number (and other) densities are not invariant. Further in F, kinetic energy (and momentum) take non-zero values, so energy, too, is transformed. A rank-2 tensor, not a simple vector, is needed to cope with both transformations (see Schutz (1985:97-8)).

Since the dust is moving relative to the frame F, there is a *flux* of particles across F-coordinate-constant surfaces. The number flux is defined surface-by-surface yielding the total flux through a volume at rest in F. **T** also incorporates these fluxes.

That yields the *spacelike* flux of particles. We can also define a *timelike* flux. Consider a t−constant (x_0−constant) surface in spacetime. Particles "cross" this surface in the sense that they endure or persist through it. If the surface is the element at that time, then the timelike flux is just the number density at the time, i.e. the number of particles per unit box at a t−constant (x_0−constant) surface.

This sketches the simplest case of the rank-2 tensor $T_{\mu\nu}$. It measures the μ component of the element's momentum across a ν−constant surface. Breaking it down into representative components (substituends of μ and ν) we have, first for the time components in each index

T^{00} − flux of x_0−momentum at a t−constant surface i.e. energy density

T^{01} − flux of energy across a x_1−constant surface

T^{10} − flux of momentum across an x_0−constant surface (momentum density)

T^{21} − flux of x_2−momentum across a x_1−constant surface

And so on through other paired indices

Note that this is a special simple example both in that dust is simple and the geometry of an element may be regarded as flat. More must be included; most notably the electromagnetic stress energy, but this introduces no importantly new features from our perspective. (See Misner et al. (1973:Chapter 5.6).) Nevertheless, the description just given in terms of the component breakdowns is a general one (Schutz, 1985:101). In dynamically more interesting fluids, the result is a more complex story about momentum and energy; for instance, fluid elements usually have random and accelerated motions. These yield important pressure terms but that does not change the broad description as given. In more complex geometries, since all closely approximate Minkowski spacetime *locally* and since the tensor is defined at points, the given breakdowns still serve locally. The tensor fields will be connected in structurally more complex ways and this poses problems for principles of conservation.

9.3.3 Pre-eminence of the metric

The field equation, just by itself, indicates no direction of explanation or dependence, being symmetrical in that respect. It can be taken as determining from the curvature of the spacetime the mass density of matter (Geroch, 1978:174). It does not exalt the right hand side as a pre-eminent source. Nevertheless that is the way it is most often regarded. Further, we can't talk about the distribution of matter or of mass-energy densities without some geometry in place. Indeed some metric or other is a necessary condition of the formulation of any matter field and of the understanding of a wide range of important concepts in GR. Lehmkuhl puts this clearly:

> We regard possession of mass-energy-momentum as an essential, rather than as an accidental, property for something to count as material. ... An object might lack the property $T_{\mu\nu}$ and still exist – but it cannot lack $T_{\mu\nu}$ and still count as a material object. ... what makes us think of systems ... as material systems is that they possess mass-energy-momentum (density). But whereas mass density was represented by a simple scalar field in Newtonian physics, relativistic mass-energy-momentum density $T_{\mu\nu}$ is defined in terms of the fundamental matter fields associated with the material system. But this is not enough: energy tensors also depend on the metric field $g_{\mu\nu}$!

Lehmkuhl (2010:464-6).

On initial value problems see Earman (1989:107).

Einstein (1983:21) was perhaps first to make the point, but in relation to the ether:

> There can be no space or part of space without gravitational potentials; for these confer on space its metrical qualities, without which it cannot be imagined at all. The existence of the gravitational field is inseparably bound up with the existence of space. On the other hand a part of space may very well be imagined without an electromagnetic field.

Thus the field equation, read at the level discussed in this section gives no reason to doubt the traditional view that **G** and **T** belong to different ontological types. In particular, it is at the very least consistent with the view that the *contents* of spacetime are exhausted in **T** fields and its *properties* in **G** fields. If that is so, then the usage that describes all tensors as geometric *objects* helps to confuse a useful, metaphysically crucial, distinction.[4] The difference between **G** and **T** embodied in the fundamental field law is well understood. The genie of spacetime should not be naturalized by absorption into dynamics.

9.4 Seeds of doubt – energy without work?

Gravitational mass and energy are widely regarded as properties of spacetime itself, to be understood as genuine instance of energy and mass. If so, there seems a reason for thinking that spacetime does belong among the entities of dynamics since the $g_{\mu\nu}$ are the components of the gravitational field. Their role as coefficients of the $d_{\mu\nu}$, the coordinate differentials, is seen as secondary. This reverses a theory whose main thrust is to reduce gravity as a force to something kinematical. That is the message of the Principle of Equivalence.

Energy is not a straightforward concept:

> It is important to realise that in physics today we have no knowledge of what energy *is*. ... there are formulas for calculating some numerical quantity, and when we add it all together it gives ... always the same number. It is an abstract thing in that it does not tell us the mechanisms or the reasons for the various formulas.
>
> Feynman (1965:vol. 1; 4-2)

I outline some rather simple reasons for thinking that the concept of energy undergoes marked and debilitating changes in GR from its robust standing in classical and SR physics. Chapter 4 demonstrated the general nature of some conceptual revision of energy in SR, although it remains a strong systematic concept there. Noether's theorems (Noether (1918), Baez (2002)) are crucially important for conservation principles in any theory of physics. Her (first) theorem states that any differential symmetry of the action of a physical system has a corresponding conservation law. The theorem was proved 1915 and published in 1918. The action of a physical system is the integral over time of a Lagrangian function from which the system's behaviour can be determined by the principle of least action. Although the theorem is confined to theories expressible in an action principle, this is not a serious limitation in this context. Where either energy or momentum must be conserved, there one needs time-translation symmetry for the first. Space translation symmetry sustains conservation of momentum in uniform motion, and, rotational symmetry sustains the conservation of angular momentum. Models of GR do not, in general, have these

---
[4]This view is expressed clearly in Anderson's influential (1967).

symmetries, yet the importance, even the meaningfulness, of energy and momentum lie in their being conserved. In the context of gravity, these apparently dynamical properties must be regarded with special caution.

Classically there are two strands of thought about what energy is. They are compatible, but raise different problems for GR. The first is about work and the definition of energy as the capacity to do it. I call this the concrete strand. The second is about its conservation and concerns summing up the various forms energy may take so that the balances among them come out constant. I call that the abstract strand, following Feynman op. cit.

9.4.1 GR is a force-reduction theory

The reductive message of GR is that gravity is not a force. Its role is taken over by spacetime structure. So, first, the classical conception of energy in terms of work done by a force fails for gravitational "action;" there is no gravitational force. Second, examination of the abstract strand shows that the principles of conservation of these new properties are different and unsatisfactory in GR. Ersatz versions of the old concepts are still very useful in many special but interesting cases. These versions are not wholly successful rewritings of the role of geometry. My main aim is to argue against the view that, in GR, the revamped versions of classical matter concepts show us that spacetime is really the gravitational field. That is a backward step. The geometry of spacetime allows us to make up something usefully like the classical concepts but in a not quite satisfactory way. Spacetime structure is always well defined in GR. Not so energy.

It was essential to the success of GR that it should reduce to something close to Newtonian physics and SR in the limiting cases of slow velocities and weak gravitational fields. It had to be able to account for the success of these theories in the past and in the appropriate limiting cases. For present purposes, this suggested that it should be natural to use the concepts and language of older theories in the context of frames of reference. These separate space from time. Natural, too, for us as experimenters, observers and theorisers to find spatial and temporal terms more intuitive than spacetime ones. But when we use the former terms, as we often do, and sometimes to advantage, it has us describing some free motions as accelerated relative to certain frames and, if we are to stick to the first law, as *forced* motions in those frames. In this conceptual environment, then, things in free fall in spacetime, may accelerate relative to $3+1$ frames and thus be accelerated by a force. This may be the product simply of coordinate choice as in Einstein's example of a rotating disk. But in others, where curvature creates the problem, one speaks of tidal force; i.e. of gravity not eliminable by mere coordinate change. Consequently, in this environment, we are obliged think of work done by gravity, and consequently have reason to write of gravitational energy – however dubious that may ultimately turn out to be. Thus there is a certain tension, or at least a complexity, in the way we find ourselves speaking of gravity, of tidal forces and of gravitational energy. That dissipates only in the conceptual context of spacetime. Perhaps that explains, to some extent, why faith in gravitational energy survives in GR – as far as it does.

Nevertheless, since GR must closely approximate Newtonian gravitation in the limit of weak gravitational fields, something behaviourally *like* energy is in the offing. We know that geometry explains certain energy and momentum changes in material systems. Perhaps one could define something – say, a potentiality for dynamical/kinematical change – as a quantitative ersatz, force-free "energy," non-dynamical energy, or the like. Then we might hope that it, plus ordinary energy, would allow an "energy" conservation principle. This turns out not quite to click. But even if it did, we would not have reduced spacetime's role to that of energy. Rather the opposite: force-free energy is understood in terms of spacetime structure, just as the force of gravity is.

9.4.2 Energy, force and work

First, a quick pass over the concept of work and the concrete strand. In elementary high school exercises on work, the core role of force is emphasised in all those problems that clarify on what work is done and by what it is done. Energy is localised. I suggest that if we deploy these skills on examples in GR we see that no work can ever be done on anything by spacetime. That follows from the arguments of Chapter 8. The classical definition of work provides no reason for speaking of energy in these contexts. That entails, in turn, that there is no gravitational energy analogous with mechanical energy, energy of heat and so on. There is again a single simple basis for saying so that stands behind every example – there is no force of gravity in GR. It is a reductive theory, eliminating gravitational force in favour of geometric structure. So spacetime exerts no force and does no work. The mechanical states of systems do indeed change in free fall in curved spacetime, or are changed by the tendency towards the free-fall of interacting constituent particles in things. The energy of a material system may change where the metric is time dependent. In curved space, momentum may change and energy change at least in form, without spacetime's doing work.

9.4.3 Conservation failure and forms of energy in curved space and spacetime

That curvature takes us into territory strange to dynamics can be seen in simple toy spaces with simple space curvature. For this, one needs no sophisticated physics. Toy spaces cropped up already in Chapter 2.2.9's use of non-Euclidean footballs and in the example in Chapter 8 §5. They make it easy to see how momentum would not be conserved. They also illustrate non-classical connections between curvature and changes in form of energy.

Consider a case of variably curved three-dimensional space. Most of it is Euclidean but suppose football-sized regions of positive and negative curvature. Let a small soft rubber ball move through a Euclidean region and run straight into one of the footballs directly towards it centre. It must change its Euclidean shape to enter the region and thus the particles that make it up will need to change their distances and orientations from one another. Elastic (electromagnetic) forces that sustain the ball's equilibrium will resist the change. Given a lucky coincidence of symmetries, the ball might be brought to rest inside the football and propelled back out again, thus first losing all its momentum and kinetic energy to elastic energy and then regaining them unchanged as it exits the football in the reverse direction.

Here again, there is no loss of time translation symmetry and no failure of conservation of energy even though its forms change. Space exerts no force on anything and changes nothing respecting any particle considered by itself. Each would traverse a geodesic of the space in inertial motion did it not interact with the electromagnetic forces brought into play by the changing distances and directions between the constituent charges inside the elastic solid.

It is worth adding that time-translation symmetry need not lapse with time curvature, even in more realistic examples. In the case of Schwarzschild spacetime, the metric coefficients for both dx^2 and dt^2 involve the factor $(1 - 2M/r)$ in different ways. Both are curvature terms, yet the metric is stationary: no time variable appears in any metric coefficient of these coordinates.

Toy cases do not tell us much about energy in GR for there is no constraint equation, no $\mathbf{G} = k\mathbf{T}$, to demand that changes in momentum and in forms of energy are associated with changes in the structure of spacetime.

9.4.4 Hard cases

More realistic examples of spacetime as gravitational energy suggest that it is best regarded as among the *contents* of spacetime rather than its structure. (I go on to disagree with each of Einstein (1916),

(1954), (1983), Rovelli (1997), Rynasiewicz (1996) (1996), Earman and Norton (1987), Norton (2011)). Gravitational radiation – gravity waves – must exist if energy is to be conserved. These waves, if they exist, will be difficult to detect far from any source and, although diligently searched for, they remain undetected. But, in principle they might be sufficiently "energetic" to destroy, say, the Rock of Gibraltar if they passed through it. That would surely make spacetime's energy spectacularly unghostly and seem to place it among familiar matter waves and fields.

> A strong burst of gravitational waves could come from the sky and knock down the rock of Gibraltar, precisely as a strong burst of electromagnetic radiation could. Why is the ... [second] "matter" and the ... [first] "space"? Why should we regard the ... [first] burst as ontologically different from the second? Clearly the distinction can now be seen as ill-founded.
>
> Rovelli (1997:193)

This misconceives the encounter of a train of gravity waves with the mighty Rock. First, consider a simpler encounter of gravitational waves with a cloud of dust – test particles at rest in an otherwise flat region of spacetime. Each particle's worldline is a geodesic. Where a pulse of gravitational waves passes through the cloud there is a rippling curvature of spacetime, a deviation of its geodesics. In the cloud's rest frame, as the pulse passes, there will be no motion of any particle in the line of the pulse's travel. But there will be motions in directions transverse to its passage. As the wave's phase varies, the particles will first move apart in one direction and together in the orthogonal transverse direction as the phase changes. But, from the spacetime perspective, each particle remains on a geodesic. However, neighbouring geodesics deviate. That projects down into what is measured in the frame (Wheeler (1990:Chapter 11)).

Each particle remains in free fall/float. None is subjected to any force, no work is done on the cloud and no energy expended on it. This follows the pattern of examples in Chapter 8.

Gibraltar is no cloud of dust but a solid lump composed of interacting particles. Each particle that makes it up will tend to trace out a spacetime geodesic, but the electromagnetic forces that make the whole a cohesive rock, strive to overcome this tendency to transverse motions and to retain the equilibrium state of the whole. *They* do work on the particles, to oppose their force-free trajectories and keep them in the equilibrium state. If the amplitude and the frequency of the waves are great enough, then, with respect to the frame, the new frame-relative acceleration of the particles could tear the equilibrium structure apart and destroy the Rock. There are certainly forces at play in this; work is done by them all on all the others. Energy is exchanged but strictly *within and among* the parts of the solid. The warping of spacetime that is the passage of the pulse through the Rock exerts no force on anything, does no work and is not the dynamical cause of its destruction. Work is done by the changing electromagnetic forces that make the solid a solid. They do work on the Rock's components and they cause its destruction. No action of spacetime shatters the Rock.

Norton gives another example – the huge energies put out by stars, of which the sun is a modest instance. Gravity initiates the compressions that give rise to this radiation. Sufficiently violent collisions may strip atoms of their outer structures and, engaging the strong force, fuse their nuclei, converting hydrogen to helium and so on.

What happens here conforms to the pattern just illustrated by gravity waves. Let's begin an idealised story, once again, substituting for the solid elastic body of the earth a gas planet of oscillating test particles (each of negligible size). Imagine these in the familiar context of a weak gravitational field such as our own. A free particle released just above the surface of the planet has a straight worldline that would take it to the centre and beyond, up to the same level at a polar point and back in a harmonic motion. (Wheeler op. cit., Chapter 4) Of course our earth particles don't oscillate like this because they immediately hit each other. The gravitational field of our fancied gas planet is created by a huge assemblage of particles collectively massive enough with their complex motions to sum to earth's mass-energy. This enables us to look at the system as entirely composed of free-fall particles doing no work and having none done on them. Their gyrations produce no heat, no radiation, no classical style energy at all, apart from the system energy of their vector sums (were spacetime simple enough to permit a meaningful vector sum).

However, if the curvature is intense and particle size not negligible there will be collisions, aggregations of matter and intense pressures as bits of matter try to follow geodesics that, relative to the rest frame of earth, crowd them together despite strong repulsive forces between particles. The result is a hot molten magma round a core of what has become a largely solid planet, Earth. In more massive cases, like the sun, there is a wholly turbulent, explosive mass that pours out energy from matter interactions that take place inside it. Spacetime plays no dynamical role whatever in this.

To have a comprehensive picture of energy that can be conserved we must, indeed, give energy and therefore mass to spacetime. But geometry is basic. The energy of spacetime is an ersatz quantity not connected to the classical foundational quantities force and work. If these examples are analysed correctly then it is false that the metric of spacetime is really energy; rather, energy is being revamped so that the metric of spacetime may be said, by courtesy, to have it.

I conclude that spacetime is really not a gravitational field in any dynamical sense. Gravitational energy is conceptually distinct from the energy of matter, matter fields and their dynamics.

9.5 Total, gravitational and T-sourced energy

9.5.1 Regional energy in curved spacetime

The differential picture of **T**, above, is defined on fluid elements. For full conservation, we need to be able to sum it over regions in which it varies from place to place. It must be "spread" to form a tensor field. This "source" energy is now to be defined for spatially extended "boxes" and the geometry of these may be very different depending on the vagaries of the frame of reference that may be chosen. Since, locally, **T** is a tensor product of vectors this requires a technique for adding vectors and tensors over variably curved regions. Adding vectors was described in Chapter 4 for the case of Minkowski spacetime where vectors add simply by adding components. It succeeds there because that spacetime is a single giant vector space. There the tangent spaces in which the vectors may be added do not differ in orientation from point to point: there are parallels.

This fails for curved spaces. There tangent spaces differ in orientation from point to point. We can add vectors and tensors only if both are in the same tangent space. A vector at a point can be added to another at a different point by parallel transporting the first vector to the vector space of the second. But the result depends on the route chosen for the transport. There is no privileged path nor, for adding several vectors at several distinct points, no privileged point to transport them to. There is no unique way to add the many vectors. Despite the definiteness of energy-momentum for single elements, the **T**-energy-momentum for a system in a curved region is not definable. In certain cases the *total* energy of a system is definite. However, in each case, this must be the sum of gravitational and matter (i.e. **T**) energy. That entails that there can be no definite gravitational energy without a well-defined **T** energy. So gravitational energy too, is indeterminate. However, there is nothing indeterminate about the geometry (See Schutz (1985:188-9)).

One should not expect the total mass-energy of a system to be the sum of the particle energies if matter exchanges its energies and momenta with spacetime itself. Further, there is what is called the gravitational self-energy of a system which is the work that would be gained by assembling the system from isolated particles at infinity. That "resides in the geometry itself and cannot be assigned to any particular particle." (Schutz loc. cit.)

The geometry and coordinates round any point at which $T^{\mu\nu}$ is defined are constrained by the very features that were used to construct it – mass and energy. Locally, in any region, an inertial frame is presupposed in specifying, differentially, the quantities in $T^{\mu\nu}$. They can't be given without presupposing that metric in the inertial frame. But in variably curved spacetimes, the metric is variably constrained in regions by variable $T^{\mu\nu}$. We don't know how to begin to analyse even dust until we have the geometry of its motions. This is a further consequence of Lehmkuhl's point that the metric is presupposed by matter specifications. They are dependent on the metric as are other quantities in GR.

9.5.2 Imposing time-translation symmetry

Hoefer (2000) notes how a total energy can be defined but only in special circumstances. He lists three, pointing out that none of them applies in the standard model of the universe.

> Integrals must be taken in the limit $r \to \infty$
> Asymptotic flatness of spacetime must be assumed
> The coordinate system must be Lorentzian asymptotically

What motivates these conditions? Are they somehow plausible or probable, their alternatives smelling non-physical? No. What recommends them is that they impose time-translation symmetry in cryptic form. They merely yield a formally necessary condition for conservation. There is nothing to be said for them independently. The apparent success is simply a contrivance. These symmetries are not present in GR models generally.

Further, the total energy of something as complex as a star (or black hole) with an orbiting satellite, is measured by tracing out the spacetime geometry round it through the track of a distant test particle (see Taylor and Wheeler (1992:Chapter 4, §§3 and 4)). Energy is approached through geometry rather than vice versa.

Time translation symmetry is guaranteed only in spacetimes where the metric is static: that is, where there is a coordinate system in which the metric coefficients contain no time variable. Many GR systems are not static in any frame of reference so no conserved energy can be defined (Schutz (1985:Chapter 7.3, 7.4)). Geometry, in contrast, is always well defined.

9.5.3 Dark energy

Astronomical observation suggests that the expansion of the universe is accelerating. In the theoretical literature, the quest is to find an energy source to generate this. Dark energy is postulated as the source. However it is unknown what it could be. What motivates this response, despite its problem, would seem to be a presumption that the solution will lie in particle physics. Yet the accelerated expansion is not different in principle from the problem Einstein believed he faced of finding a static model for GR where change is just what the simplest form of the equation predicts. He introduced a cosmological constant to achieve this. It is a scalar coefficient of the metric, thus added to the *geometry* side of the equation. There is no particular reason to assume that it must lie in augmentation of **T**. If gravitational energy really is reducible to spacetime geometry then dark energy may not be needed. One might look instead for a spacetime structure that, of itself, yields an accelerated expansion. De Sitter spacetime suggests a direction in which a search for this might begin. But de Sitter is a vacuum solution to the field equation.

None of this makes gravitational energy useless as a concept in GR models. But it is essentially a rewriting of the metric and it does not lie at the foundations of GR.

9.6 Conservation problems

Feynman (loc. cit.) makes it clear that energy must be conserved if it is to have either a fundamental or a leading role among the basic concepts of GR. It is not clear how a quantity or a system of quantities could be regarded as energy unless this condition is met. The requirement is not satisfied in GR generally.

A brief description of **T** was given for the differential case in §3.2. Define a small volume (box) in spacetime and a surface containing it. The energy in the volume is conserved if any loss of energy through the surface is balanced by equal gains across it. Formally, a theorem of Gauss can be used to define the volume and the surface and the total flux. If energy is conserved, this total should be zero. Then total energy remains the same within the volume.

We can consider a volume with pairs of parallel coordinate surfaces (including $t-$constant surfaces) containing it. In SR, at least, the *divergence* of **T** sums the flux of energy across all these surfaces and so determines the net flow of energy and momentum into and out of the volume. If, and only if, the divergence is 0 (vanishes) then energy and momentum are conserved. Formally, the divergence sums the gradients of **T** across each surface. It is written

$$T^{\mu\nu}{}_{,\nu}$$

SR spacetime has the symmetries needed for all standard forms of conservation but none is a general feature of models in GR. We saw in Chapter 4 that spacetime plays a dominant role in defining the conservation of energy and momentum. But that rests on symmetries not available in curved spacetimes. Conservation calls for an analogue of this divergence in GR.

The Einstein tensor $G^{\mu\nu}$ was constructed with the intention that its covariant divergence will automatically vanish. So the covariant divergence of the matter tensor $T^{\mu\nu}{}_{;\nu}$ must vanish, too, to satisfy the field equation. As Brown (2005:141) puts it "This is about as close as anything is in GR to the statement of a conservation principle." Hoefer (2000) pursues, in detail, the theme that, even so, its vanishing does not really yield energy conservation. Further, as he illustrates from the literature, there is not just some attitudinal ambivalence about its adequacy: there is downright inconsistency. It is frequently affirmed, and no less often denied, to be a genuine conservation principle – this in one and the same publication, including in Einstein's original paper (see Einstein (1916:151)).

$T^{\mu\nu}{}_{;\nu}$ is written with a ";" since its expansion is more complex than the "," or ordinary divergence:

$$T^{\mu\nu}{}_{,\nu} + \Gamma^{\mu}{}_{\alpha\nu}T^{\alpha\nu} + \Gamma^{\nu}{}_{\alpha\nu}T^{\mu\alpha}$$

The ordinary divergence is just the first term of this sum. The two Γ terms are related to the coordinates and occur whenever they depart from simple Cartesian or Lorentzian form. For instance, if one chooses polar instead of Lorentzian coordinates, one needs these terms even in SR. But, in SR, Γ terms can always be made to vanish by return to Lorentzian coordinates. But not in GR since the curvatures in most spacetimes forbids their use. Γs are there because the spacetime structure itself demands them. If that structure is to be construed somehow as energy, Γs are highly relevant additions to the ordinary divergence.

The application of Gauss's Law to the ordinary divergence involves straightforward integration over its single term. But this is not possible with the complex of terms in the covariant divergence.

One strategy at this point is to invoke the "comma to semicolon rule." It rests on the Principle of Equivalence read as entailing that a local inertial frame is available at any point in any GR spacetime.[5] On that basis the ordinary divergence is equivalent *locally*, if expressed in local Lorentz coordinates, to the covariant divergence. The Strong Principle of Equivalence states:

---

[5]Einstein always stated the Principle as quoted in §9.2.2 above.

> Any physical law which can be expressed in tensor notation in SR has exactly the same form in a locally inertial frame of a curved spacetime.
>
> Schutz (1985:184)

This applies in the tangent space at any point. It is suggested, further, that one can validly move from one tangent space to a neighbouring one. This was advanced by Einstein and Fokker in an early paper (Einstein and Fokker, 1914).

> Einstein and Fokker's assertion ... that the laws of the original Theory of Relativity can be matched in the new generalized theory amounts to a bold but not implausible hypothesis that the generally covariant equations which, on small neighbourhoods, link the local fields to one another, hold also in the same form between the respective global fields.
>
> Torretti (1983:157)

But while this version of the Strong principle is intuitive in the limit, it is not consistent with "curvature coupling" (Schutz (1985:184), Hoefer (2000:192)). The Riemann tensor must play no role in the formulation of Maxwell's equations for GR. No such coupling has been observed (in our world, of course), but it is alien to the fundamental principles of GR that curvature may simply be neglected in this way in all GR-possible worlds.

Another approach to the problem stresses that T is constructed, as we saw above, to exhaust the energies constituted by matter fields – the only energy that relationists regard as real. It does not include gravitational energy. That **T** is not conserved by itself should come as no surprise, as Einstein noted from the start. The *ordinary* divergence of **T** shouldn't vanish, because of "energy exchanges" with matter and because there may be gravitational inputs (e.g. waves).

This has suggested a different approach, that the Γ connection terms could be reformulated as a gravitational energy tensor, usually written as **t**, definable through derivatives of the metric tensor. The ordinary divergence of the sum of the two could then seem a proper principle of conservation, formulated as $(T^{\mu\nu} + t^{\mu\nu})_{,\nu} = 0$. The use of that expression in Gauss's law might then be an acceptable conservation principle.

The rewriting is possible but ugly and there are infinitely many distinct ways to do it. That looks uncomfortably arbitrary, but not trivial. If **t** is to represent a tangible physical-looking quantity, then it ought to be a tensor, a covariant object. It is not. Adroit shuffling of coordinates can always make it vanish at any point, so that this gravitational energy is not satisfactorily local so as to tie it properly to *work*. Hoefer (op. cit.) concludes that energy in GR is dubious. I agree.

The **t** pseudo-tensor does not fit the classical idea of energy as work done by a force because "gravitational energy-momentum" is not localisable. In any small region of spacetime an appropriate choice of reference frame will make it 0 since there is no such thing as local gravitational energy-momentum. That is guaranteed by the Equivalence Principle. What **t** should capture is a global effect, not a local one. As Misner et al. (1973) put it, "The overall effect one is looking for is a global effect not a local effect. That is what the mathematics cries out. That is the lesson of the non-uniqueness of the $t^{\mu\nu}$!" (op. cit. 468)

My aim in this chapter is to argue that the claims made for ridding GR of the genie by claiming that it is, after all, familiar dynamics, do not succeed. Geometric structure is ontologically unique and, in its explanatory role, unprecedented, despite being uncomfortably unfamiliar in its metaphysics. It has not been my (impertinent) aim to persuade physicists that they should never speak of gravitational energy. It is useful given the key role that energy played in classical and SR physics and the elegance and power that it bestows in the study of many models. However, I do mean to persuade philosophers that GR does not justify construing spacetime as energy. Rather it was always, and remains fundamentally, a theory that construes gravitational energy as a structure of spacetime.

10 THE TROUBLE WITH GENERAL COVARIANCE

Chapter 10 tackles an objection to substantivalism called the Hole Argument which has held a dominant place in the literature since its first version in Earman and Norton (1987). The chapter begins with an account of GR models or worlds that are possible if GR is true. It gives a brief introduction to the differential geometry crucial to the construction of models. There is just enough exposition to float the Hole Argument and the following critique of it. The Hole Argument itself is introduced by means of a simple example at no cost in generality. Lastly the argument is examined critically and found wanting in a number of respects. In particular it is argued that it has no proper claim to be a metaphysical argument.

10.1 Introduction

In his Autobiography Einstein wrote:

> In a gravitational field (of small spatial extension) things behave as they do in a space free of gravitation... This happened in 1908. Why were another seven years required for the construction of the general theory of relativity? The main reason lies in the fact that it is not so easy to free oneself from the idea that coordinates must have an immediate metrical meaning.
>
> Einstein in Schilpp (1949:67)

His problem was how to *formulate* GR. Its solution is to write the theory in a *generally covariant* style. Einstein (1916:§3) placed much ontological weight on this and other features of the new theory.

> That this requirement of general covariance, which takes away from space and time the last remnant of physical ob-

jectivity, is a natural one, will be seen from the following reflexion. All our space-time verifications invariably amount to a determination of space-time coincidences... Moreover, the results of our measurings are nothing but verifications of the meetings of the material points of our measuring instruments with other material points, coincidences between the hands of a clock and points on a clock dial, and observed point-event happenings at the same place at the same time. ... As all our physical experience can be reduced to such coincidences, there is no immediate reason for preferring certain systems of coordinates to others, that is to say, we arrive at the requirement of general covariance.

(Op. cit. 117-118)

All of this was mistaken. What most needs to be explained is why adopting the generally covariant style of formulation entails little about the structure of space, time and spacetime. That, and broader issues of general covariance, are the theme of this chapter. What emerges is a remarkable instance of metaphysical error leading the most eminent of scientists astray.

10.2 General covariance: an introduction[1]

Readers familiar with this topic may wish to skip.

10.2.1 The case of the Euclidean plane

Classical mechanics from Newton on assumed that space must be Euclidean. That geometry brings with it the huge advantage that one can use Cartesian coordinates. They yield a quite simple *special covariance*. Once understood, the move from special to general covariance is easy. For simplicity, I'll stick for some time with Euclidean geometry in 2 dimensions.

Cartesians are easy because they exploit the parallels structure of Euclidean space which also enables orthogonal intersections for x and y coordinate lines. Also with Euclid, the metrical structure is the same everywhere in the space and at every time. That yields, everywhere,

---

[1] Norton (1992:§5.4: 195-203) is strongly recommended for readers unfamiliar with these ideas. In Norton's example pp. 201-2 the covector or one-form, dT, is a simple tensor and a covariant of the general group of transformations. For more detail see Kretschmann (1917), Anderson (1967), Torretti (1978), Norton (1988), (1992), Earman (1973), (2006).

the familiar uniform rectangular coordinate grid. Then you can assign coordinate *numbers* to points so that, for instance, a difference in the coordinate numbers of two points on the x coordinate line, encodes a distance in *space* between the points, so long as you have standards and units of measurement that allow the number difference to represent the number of metres of distance between the points themselves. Minkowski spacetime has a similar advantage for similar reasons in the use of Lorentz coordinates. It will be enough to look just at the Euclidean case.

Consider a distance in Euclidean 2-space between any two points p and q. A Pythagorean function of their coordinate differences yields the distance in a difference equation:

$$\Delta s^2 = \Delta x^2 + \Delta y^2$$

That is very like the crucially significant metric tensor for this space. Its *components* are the pair of coordinate differences Δx and Δy.

There are infinitely many distinct systems of Cartesian coordinates. Three basic transformations map any of them to any other: translations (shift the origin in any direction) rotations and reflections. Each is described by a simple algebraic transformation equation that tells how the xy system is transformed into the $x'y'$ system. The relevant equation tells you how the coordinates of the distant endpoints change and thereby how the components change. The components are *covariants* of the transformation, their instances varying by the same equation as the system transformation does. So instances of the equation are covariant too. The Pythagorean function is exploited to deliver the distance as before but from the new components. Obviously it will be the same distance in the new coordinates so Δs^2 is an *invariant* of the transformation. This sort of covariance is *special* to the specially simple Cartesians. In classical physics many interesting expressions are special (Cartesian) covariants including the laws of mechanics themselves.

Now for a step towards *general* covariance. Since coordinates may be freely chosen one need not choose Cartesians. It may be convenient to use polar coordinates. Then the covariant structures are more complex. This is most easily seen in skew coordinates. These resemble Cartesians save for jettisoning the advantage of orthogonal intersection of x and y coordinate lines in favour of acute angle (θ) intersection. Then a more complex function yields the distance from its components:

$$\Delta s^2 = \Delta x^2 + 2\cos\theta\,\Delta x \Delta y + \Delta y^2$$

Information that was encoded in Cartesians now becomes explicit in the coefficients of the cross terms in the distance equation rather than in the simple coordinate differences themselves.

Expanding this equation clarifies a more general and basic algebraic structure underlying this:

$$\Delta s^2 = 1\,\Delta x \Delta x + \cos\theta\,\Delta x \Delta y + \cos\theta\,\Delta y \Delta x + 1\,\Delta y \Delta y$$

This is the sum of all 4 forms of product of the coordinate differences, each with a coefficient (in this special case the coefficients are 1 and $\cos\theta$). Only the coefficients, written as "g" with subscripts, vary as the coordinates change. To make our notation more elegant and general change "x" and "y" to "x_1" and "x_2." The general expression for distance in 2 dimensions then becomes the neat:

$$\Delta s^2 = g_{11}\,\Delta x_1 \Delta x_1 + g_{12}\,\Delta x_1 \Delta x_2 + g_{21}\,\Delta x_2 \Delta x_1 + g_{22}\,\Delta x_2 \Delta x_2$$

Neater still, record just the 4 coefficients in a square array:

$$g_{11} \quad\quad g_{12}$$

$$g_{21} \quad\quad g_{22}$$

The index of a coefficient assigns it to particular product of coordinate differences and these products may be omitted since they occur in the same pattern in every instance of the equation for Δs^2.

Arbitrarily curved coordinates may be used. The coefficients in the metric then become quite complex. Generally, coordinate curves need be neither straight nor parallel nor intersect orthogonally. Attempts to encode differences in coordinate numbers to the size of distances in space then become impossible. There is a constraint on these coordinates nevertheless. It is described in the next paragraph. The resulting coordinates are *general*. Coordinate number differences no longer encode the number of units of distance in space itself since the curves may vary wildly. The equations that yield distances and define the metric will differ from point to point in terms just of the coefficients. The metric in such coordinates must be given point by point.

Whether coordinates are Cartesian or general is not simply a matter of how quadruples are assigned to numbers. More important is which group of equations are envisaged as transforming one coordinate system into another. Cartesian coordinates are subject to a small group of transformations, the Euclidean group. Those transformations take us from one Cartesian system to another. It is the smallness of that group that allows us to read them e.g. as encoding distances in space as coordinate-number differences. The most general group erases this and much else that Cartesians encode, since its transformations carry us from one system to another by any transformation that is continuous and differentiable. These transformations do not erase the coordinate encoding of all spatial structure. The smooth continuous ordering of the quadruples still captures the smooth continuous structure of the space – its differential topology in short. Contrary to Einstein's remarks above, while that structure is perhaps "the last remnant of physical objectivity" that belongs to space and time, general covariance does not take it away but *preserves* that crucial remnant of spatiotemporality. It preserves the relations of connection and separation that are the foundation of spacetime realism.

We are now at the problem that was "not so easy" for Einstein. For very small distances where smooth coordinate lines don't vary much from straight, distance magnitudes are derived from the same basic product of coordinate differential components and with coefficients that record how the components are to be combined in that small region so as to yield the distance. The general expression for the line element in calculus notation now is:

$$ds^2 = g_{11}\,dx_1 dx_1 + g_{12}\,dx_1 dx_2 + g_{21}\,dx_2 dx_1 + g_{22}\,dx_2 dx_2$$

This is the *metric tensor* in the chosen coordinates. It is succinctly expressed as g_{ik}, with i and k indices ranging over the coordinates in the dimensions required for a given space or spacetime. In GR spacetimes there are 4 dimension and so 4 coordinates for each point. Then the tensor is a 4×4 array of 16 components not all of which are independent. For however many dimensions and for any Riemannian space (one in which the metric tensor has the quadratic form of the last equation above) the tensor may be written simply as g_{ik} where the indices i and k take each of the coordinates in turn as values thus pairing the coefficients with the relevant differential products.

Tensors loom large in GR. They are complexes of vectors which are, themselves, the simplest tensors. Vectors have direction and magnitude. Assuming Euclidean space and Cartesian coordinates they can be diagrammed as arrows. They represent direction and, magnitude, pictured by its length. That represents the vector bilocally. It has distinct endpoints.

But few vectors have real bi-locality. An electric vector has direction and magnitude but it locates the field strength and direction only at its butt end. Its magnitude and direction are not intended to entail anything about the field at the sharp end. So vectors really inform us only about what is at one point.

This is also an issue when we represent vectors in coordinates. The coordinates refer us to a point and the vector components tell us about the magnitude and direction of something that is there. If the space and the coordinates are not Euclidean or Cartesian so that the geometric and coordinate structures change from point to point in ways constrained only by continuity and differentiability, then we can't represent them as arrows reaching out into arbitrary spaces in arbitrary coordinates. Nor can we assume that the familiar operations of vector addition, multiplication and so on will yield determinate results in these wayward circumstances.

Each Riemannian space approaches flatness in some sufficiently small region round each point so that we can envisage a Euclidean (Minkowskian) space tangent to it at the point. The vectors and tensors are conceived of as residing in this tangent space as a vector space of unlimited size. Then vector components define direction and magnitude in that ideal space and in that context operations on them conform to familiar vector algebra. The tangent spaces are then oriented towards each other in physical space by means of an affine connection. We need not worry about just what that is. It also defines parallel transport of vectors and tensors.

Formally a Euclidean 2−vector's magnitude is calculated from its coordinate components by the same equation as for a distance. But the direction of the vector is lost without the signs of its components. So it is regarded, not as the condition in space at the point where it resides, but as the couple of its components. Whereas the distance is a simple invariant of the coordinate transformation its direction is captured partly in their signs. The vector is then the couple of its components in the coordinates in use. So the vector is not strictly the magnitude and direction of some entity in the space but the covariant pair of components. Sometimes vectors and tensors are spoken of as the absolute objects in spacetime and as independent of coordinates.

Since the coordinates are now general a theory formulated in them is called *generally covariant*. Its laws will be formulated so as to be invariants of the transformations of the general group. The take-away message of this section is that, when a general coordinate formulation of a theory is used geometrical information that could be encoded in the coordinates as Cartesians (and others) do, it is not lost but stored in the metric (or other) tensor field.

If the geometry of the space is not Euclidean (or Minkowskian for spacetime) it will forbid any use of Cartesian or Lorentz coordinates. In that case the more complex metric tensor is explicit from the start.

10.2.2 The case of GR

Einstein's route to general covariance was not through considering arbitrary choices among coordinates. If it had been it seems unlikely that he would have made the error noted by Kretschmann. He began with the problem how to formulate a theory in such a way as to leave its fundamental differential equation committed to no more geometry than would be common to all models of his theory. No one had addressed such a problem before. The problems of variation in spacetime structures and (for quite different reasons) variation in coordinates were entwined for him.

The complexities of formulation discussed in the last section arise simply because coordinates can be freely chosen. The structure of the space they are to describe has not been an issue. Let's expand our view.

Cartesian coordinates are available only if space is Euclidean. Then, at each point in the 2D plane we have been considering, just one line is parallel to, say, the coordinate line through the origin of the coordinates. Only Euclidean space has parallels, so the Cartesian demand for them can be met in no other space. In the range of Riemannian spaces, Cartesian coordinates with their "immediate metrical meaning" are not generally available. Indeed, many spacetimes of interest do not have even Euclidean topology although one can always analyse them into Euclidean subregions that can be smoothly pieced together, like the pages of an atlas. Each subregion is Euclidean in its topology so they can be joined smoothly together (with overlaps etc.) to form the topological equivalent of, e.g., the surface of a sphere or a torus. Before GR every theory assumed the same uniform structures of space and its time in all its models. To some extent, GR can do the same but only if the assumed structure is weaker than metrical. The structure common to all models is a merely topological one, called a differentiable manifold, often simply "manifold." Its structure consisted in just relations of separation and connection that are continuous and differentiable (smooth). That is rash realism in the nomenclature of Chapter 2.

The metrical assumption underlay not just the simplicity but also the determinism of earlier theories since the laws themselves presupposed Euclidean geometry specifically. Once initial conditions were given in these coordinates, it was straightforward to deduce the positions and times of consequences numerically in the same coordinates. Granted that the space admits Cartesian or Lorentzian coordinates, assigning them to any one spatiotemporal region fixes them throughout space and time – they automatically extend infinitely and are defined everywhere. The ubiquitous metric and the elegant encoding of it in infinitely extendible Cartesian coordinates, met the requirements of prediction simply and precisely.

A final remark will complete this thumbnail sketch of coordinates and covariance. A differentiable manifold is called a topological space sharing its smooth topology with that of Euclidean space. Its coordinates encode that bare structure (in subregions of the space) but no richer ones. But a $1-1$ correspondence and a general transformation group *represent* or *model* a topological space. Geometers often speak of a space when they mean the imaginary space of a possible world, not our physical space. There are "spaces" of constant curvature or of 17 dimensions but they are only mathematical entities. "Space" can mean a purely mathematical object, abstract, fictional, unreal. Yet if GR is true, our physical spacetime is a real space and also differentiable manifold just as a real warthog is also an animal. This does not entail that our spacetime is devoid of all but differential structure any more than warthog's being animal implies that it somehow manages to be real without the properties that distinguish it from a tiger. Later we will find that realists and substantivalists are strongly advised, nevertheless, to take our spacetime as being *only* a differentiable manifold devoid of richer properties. It pays to be alive to the senses of "space" at the start of these considerations.

10.3 Theories, models and worlds

The broad *theory* of GR can be formulated as an equation only in general coordinates. The fundamental field equation:

$$G_{\mu\nu} = k\, T_{\mu\nu}$$

can't be written so as to presuppose any one metric spacetime since it tells us how the metric *differs* relative to the *different* mass-energies and momenta in various models and to the varying matter distribution within any particular model. It is *solutions* of this differential equation that provide models (GR possible worlds) with definite metrics and mass-energy backgrounds. The manifold and the general covariance group identify a common basis on which any model may be built. The manifold gives a common perspective on the structure of all models and a basis for comparing widely different ones.

Most work in General Relativity explores one or other solution of the fundamental equation and thereby a possible world or local sets of conditions. These seldom feature the differentiable manifold. Boundary conditions and initial values may be postulated in order to determine solutions of the fundamental law. The first exact solution of Einstein's equation was for the simple but important general case of a single non-rotating spherical mass, something like the sun, with a Minkowski metric at infinity. It was found by Karl Schwarzschild in 1916, the same year as Einstein published the theory. He arrived at both a metric and specific coordinates for it. They are named after him. In this solution, the metric tensor is everywhere quite simply expressed in these coordinates. The metric is indeed variable but, in these coordinates, it is a function just of r, the radial interval from the centre of symmetry. This is the starting point for studying black holes, including spinning and imperfectly symmetrical ones.

As we saw, in the special solution that is Minkowski spacetime, a metric can fully encode spacetime intervals in Lorentz coordinates. These coordinates permit the special Lorentz group of transformations with all its advantages globally. But such simple solutions are special indeed and often the first problem is to find coordinates and a metric tensor that are workable in the case to be explored.

Every theory of mechanics and electrodynamics earlier than GR can be formulated in a generally covariant style. No earlier theory makes that style attractive. That GR *can* be formulated in generally covariant fashion tells us nothing about the world not already entailed by earlier theories. That GR *must* be so formulated if all models are to have a common basis is significant. It means that Euclidean and Minkowskian geometries are dislodged from their privileged place in favour of the overarching class of Riemannian geometries. It also tells us that while our spacetime has at least the structure of a differentiable manifold it also has a richer array of properties up to metrical ones. Discovering that every classical space, time and spacetime is a differentiable manifold is like discovering that all the things studied in zoology are animals. It is basic, true, but uninformative.

The fundamental equation of GR places mutual constraints on both geometrical and matter structures that forbid the older, easier way of model building. We can't start in the old way since we don't know which metric spacetime, among an infinity of possible ones, is ours. Even if we did know which one locally, we know that, in general, it won't be the same globally.

This is not because there isn't a metric spacetime in each GR world. Every model, every global solution of the fundamental equation, has one. The universe we live in, supposing GR true of it, has exactly one, particular, complex, variably curved, metric spacetime. In some sense or other it contains everything else; in some sense or other it is a background. All geometric tensor structures, including those weaker than the metric, are similarly constrained by the **T** tensor in the fundamental equation: e.g. affine, projective and conformal structures are. The manifold encodes no geometric structure that falls within the scope of the field equation. It does not even distinguish space from time although it is four-dimensional. Its differential geometry is in no way constrained by matter. That is why it is fundamental.

Despite its weakness, the manifold may be too strong. We do not know whether spacetime is continuous and differentiable or whether it really has only four dimensions. That is not the concern of this book. But it should be understood that the manifold is postulated as continuous and smooth largely because the standard mathematical tools of physics, differential equations, are unusable without assuming continuity and smoothness. The assumption looks far less plausible than it did before the 20th century. But it is not known to be false and it is better understood than its rivals.

Knowing about the manifold, the general covariance group and the role of geometric tensors is not knowing how anything in the world *works*. It is not theoretical knowledge but an understanding merely of how theories may be formulated.

GR entails that, in any world, matter and fields are distributed in a spacetime with a pseudo-Riemannian metric. Our world can be described on the basis of a manifold, by geometric and matter tensors on it mutually constrained by the fundamental field equation. It is standard practice to denote the broad structure of a solution of the fundamental equation, i.e. a model, as an ordered triple:

$$< M, \mathbf{g}, \mathbf{T} >$$

where M is the manifold, $\mathbf{g}$ the metric field and $\mathbf{T}$ the matter field.

One prime motivator of a relationist approach to spacetime ontology makes little sense in GR models. None of the Leibniz shifts is a symmetry of variably curved spaces or spacetimes, as we saw in Chapter 1. The shifts can't even be properly formulated. The diffeomorphisms that set up the problem of the Hole Argument of the next sections §5.3 and §5.4 are symmetries only of the manifold, not generally of metric spacetime.

10.4 Approaching the Hole Argument

10.4.1 Active and passive transformations

So far, general covariance has been about transformations of coordinates under the general group. These are called *passive* transformations since, while coordinates change, nothing in the physical world changes. *Active* transformations (so called) of a space or spacetime envisage comparisons of structure in one region of spacetime with that in other regions. The comparison is effected by *mapping* regions of a spacetime onto others, resulting in matches (or failures of match) of structure in the regions.

How is this active? A mapping is analogous to, and sometimes called, a motion. Euclidean space is highly symmetrical and its metric the same in all regions. This entails that if we can imagine something like moving any one region in space across to any other we can compare them directly and show their metrics to be the same. Such an operation will be a symmetry of the space and the metric an invariant of it. Obviously neither the space nor any region of it can literally be moved. They are called automorphisms to distinguish them from maps which take us from one space to an entirely different space. So "motion" and "active" are metaphors for abstract operations, mappings of the regions into regions or the space into itself. Obviously the mapping (motion) must not alter the properties to be compared. In Euclidean space, a "rigid motion" of the space across itself means that the operation does not change the metric structure.

General "active" transformations of any space use the group of *diffeomorphisms*, the continuous and smooth "motions" of space into itself. They preserve the structure of continuous differentiability and thus are symmetries of the topology of the space. So it is the manifold structure, M, that is invariant under the general diffeomorphic group. One distinguishes M from a diffeomorphism of it, $\mathbf{d}(M)$, since the points of M are viewed as mapped to other points of it. What is at issue in active general covariance is whether or not spacetime has the same differentiable structure in every region. Diffeomorphisms compare only that weak level of structure and the symmetry is therefore rather a trivial one. Analogously with the Euclidean case, the result is a symmetry and the relations of smooth connection and separation will be invariant.

The mappings envisaged in active *general* covariance are formally the same as the *general* group of coordinate transformations. In GR, just as the laws of the theory are invariant under the general group of coordinate mappings, they are also invariant under the general group of active mappings. Norton (2011:§2) puts it as follows: in "active general covariance we are licensed to spread geometrical structures like metric fields over the manifold in as many different ways as there are coordinate transformations."

10.4.2 Leibniz shifts vs diffeomorphisms

Active (in some sense) though the Leibniz shifts are, they are not like these transformation. For one thing we can imagine concrete operations corresponding to them. God's ways may be mysterious but need not be abstract. In a Leibniz static shift, the whole *material* world is shifted or mapped, all of it together, to new locations in space. It *detaches* the things from their relations to space, translates them, then *re-attaches* them. The active transformations of the manifold are not shifts of the *contents* of space but shifts of the space itself: it's as if they copy one region of space, (spacetime) move the copy unchanged to another region, then directly compare the weak geometries of the copy and the new region.

A distinction between Leibniz shifts and active transformations may seem pedantic. But not if our purpose is to invoke the *metaphysical principle* used by Leibniz to draw the conclusion that spatial relations among things are independent of their relations to space.

Leaving aside theological principles, the Principle of the Identity of Indiscernibles, the inspiration for Leibniz Equivalence, is the relevant thesis here. To invoke that correctly in this context needs a specific form of the detachment thesis. A Leibniz shift or mapping *detaches* the *objects* from their spatial relations and replaces them in other locations with *structurally* similar but *individually* different relations to space. In an active transformation, it's the *spatial region* that is mapped to another. Objects in a region remain *attached* to the mapped points; they *tag along* with them in a sense soon to be explained more carefully. Active transformations are like the holus-bolus shifts mentioned in Chapter 1.4. They are a different kettle of fish from Leibniz shifts.

What happens to objects in an active transformation? Recall the details of a generally covariant formulation of GR. Since the covariance group of interest will be the general group, any relevant model has the structure $< M, \mathbf{g}, \mathbf{T} >$. Diffeomorphic transformations change the manifold but not its smooth structure. Nor do they touch anything intrinsic to the geometric and the matter tensor fields, $\mathbf{g}$ and $\mathbf{T}$. Points of the manifold are mapped and their attached tensors, *both matter and geometric*, follow them unchanged. They represent, at new manifold points, just what they did at the original points.

10.4.3 On formulation and determinism

The Hole Argument is relevant only to the debatable view that the differentiable manifold's weak structure best represents a GR spacetime (Norton, 2011:§4). It presupposes interpretations of the fields of GR like those examined in the last chapter. It states that the projective, conformal, affine and metric structures, expressed in appropriate tensors, are not *properties* of spacetime but among its contents just as the dynamical or matter tensors are.

The substantivalist is urged to accept this presupposition. How should we try to understand it? No substantivalist would accept spacetime as an *abstract mathematical object* since that is not a substance. The spacetime we live, move and breathe in – not a merely possible world but our concrete real one – certainly has at least the structure of a differentiable manifold but it is not an abstract mathematical object. What is urged on the substantivalist is to take our spacetime as having manifold structure as its *only* property. But it makes no plain sense that there are animals that lack any properties of the kind that differentiate warthogs, tigers and people. Just so, it is not obvious that real spacetime can have only that meagre structure. On the face of it nothing is just an animal and nothing just a manifold.

Two closely related problems are raised by general covariance:[2]

---

[2]Earman and Norton (1987:§1 and §5).

first, if the manifold is an adequate representation of spacetime then there are infinitely many indistinguishable spacetime models; second, this robs us of the power to predict at least one aspect of every state of GR affairs viz. which model it is in. The latter objection has been more prominent. This dominated the literature on spacetime during the 90s and is widely taken even among non-specialists (see e.g. Ladyman (2009) 4.3) as showing decisively that the price of substantivalist's view of spacetime is a failure of determinism.

Some GR spacetimes are causally ill-behaved in not being time-orientable: i.e, no time-directed vector field is continuous over the whole of spacetime. Gödel's well-known model is an example.[3] For determinism one needs a spacetime with a Cauchy surface (roughly a spacelike hypersurface that may figure as an instant of time throughout the model's space relative to some coordinate system). If the model has such a surface then initial values may be defined on it that yield a determined future (and past).[4] However, the objection from many indistinguishable models is independent of this style of determinism.

10.5 The Hole Argument[5]

It will serve my purpose to explore the Hole Argument for indeterminism in a model set in Minkowski spacetime where one certainly expects determinism. I will argue that it fails in that simplest case and thereby fails in general. As Earman and Norton have stressed, the Hole Argument applies to all theories of mechanics that precede GR and does not depend on curvatures in spacetime.

Suppose a model $< M, \mathbf{g}, \mathbf{T} >$ where M is its manifold, $\mathbf{g}$ its metric tensor field yielding a Minkowski spacetime, $\mathbf{T}$ its matter tensor field composed of test particles (i.e $\mathbf{T}$ is everywhere virtually 0 yet we can speak, illustratively anywhere, of particles). Let $\mathbf{S}$ be a spacelike hypersurface, dividing spacetime into two temporal regions (i.e. a Cauchy surface). A proper subregion, R, of M includes $\mathbf{S}$ and all of spacetime preceding it. The Hole in this example is all the rest of spacetime later than $\mathbf{S}$. Let C be a particular Lorentz coordinate system for the whole spacetime; suppose there is a valid and sound prediction, P, in C coordinates, from initial conditions all of which lie within R. P predicts an emission event, e, of a photon from a particular source; e lies beyond R and in the Hole.

---

[3] Gödel (1949); Malament (1986).

[4] For a rigorous account see Wald (1984:Chapter 8, especially §8.2).

[5] The literature on this argument is voluminous. For a sample see Brighouse (1994) (1994), Butterfield (1987), (1988), (1989), Earman (1973), (2006), Maudlin (1990), Norton (1988), Rynasiewicz (1994).

Now consider a diffeomorphism, **d**, of M. It maps each point of M in the region R onto itself, but maps each point, p, in the Hole (each point in the region beyond R) smoothly onto a different point **d**(p). In this diffeomorphism the tensors in the fields **g** and **T**, together with e with all its properties, follow the points they are attached to in the mapping. We let the coordinates follow too. This creates *a new, a distinct GR model*. It is an active change (motion) in the manifold so that e is mapped with p to a different manifold point **d**(p). The prediction P still foretells e, the emission of a photon from a particular source. The prediction, like the **g** and **T** fields, is a diffeomorphic invariant. Beyond **S**, thus in the Hole, **d** has spread the tensor fields over the manifold in a new way. What cannot be predicted – not from all the information in R i.e. all the history of this possible world preceding **S** – is which manifold point e is located at. The distinct models generated by **d** are nevertheless indistinguishable in terms of any properties of the **g** and **T** fields.

So if substantivalism's best bet is to identify spacetime with the bare differentiable manifold then it is lumbered with an indeterministic theory before any serious issue about determinism arises. So it turns out to be the worst bet. If we take any richer structure to be what spacetime is then its structure is not invariant under diffeomorphisms and the problem does not arise.

10.6 Critique of the Hole Argument

Despite all this, the information available in R allows strong predictions about the states of affairs in the Hole – all you ever wanted to know.[6] The information is available in every diffeomorphism). It lies in the tensors and tensor fields attached by coordinate quadruples to each manifold point so that all this tags along untouched by the diffeomorphism. In the example, whether we consider M or **d**(M), that tagged along information entails that all of spacetime, including The Hole, has a Minkowskian metric. It also entails that, on any **d**(M), the coordinate system may be taken as Lorentzian. It transforms under the Lorentz group. The coordinates encode the information needed. An appropriate calculation correctly predicts both the magnitude of the spacetime interval and the 4-vecor orientation between any event in R and the photon emission event. These are determined by the metric field and are independent of which manifold point is relevant.

---

[6]Determinism is more difficult than this discussion takes account of. For instance, even in classical physics, information from objects and fields may rush in from arbitrary distances and swamp prediction's boats. This aspect of threat to determinism is left aside so as to focus just on what flows from general covariance. See Earman (1986).

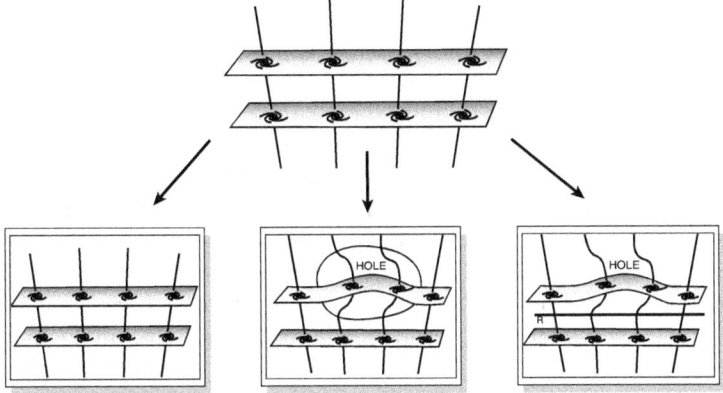

Figure 10.1: The top diagram represents the Minkowski spacetime of the world discussed in the text. Its contents are sketched by the worldlines of its test particles and their intersections with two Cauchy surfaces. Each diagram below represents a coordinate description of the world. The left diagram shows it as Lorentz coordinates do. The central diagram shows one version of a Hole, a version preferred by Norton (2011). It is a diffeomorphism of the left diagram yielding an identity map outside the closed curve (the Hole). Inside it each point is mapped to a distinct point (as suggested by the warping of the worldlines and surfaces). The right diagram represents the case discussed in the text where the diffeomorphism in an identity on R and all before it. Beyond R each point is mapped to a distinct point (as suggested again in the warping of the worldlines and surfaces). The diagram is suggestive rather than precise since it is a metrical representation of topological structures.

The only thing we don't know is e's location on the manifold $\mathbf{d}(M)$.

10.6.1 The models really are distinct

How all this works out can be most briefly seen in the standard model structure $< M, \mathbf{g}, \mathbf{T} >$. Predictions all go on at the $\mathbf{g}$ and $\mathbf{T}$ levels. M provides the smooth topology that spreads tensor fields; it provides the indispensible separation and connection relations and the smoothness. It provides nothing more and, given its poverty of structure, can provide nothing more. So which tensors are assigned to which points and which fields to which regions is necessarily arbitrary up to a diffeomorphism. $\mathbf{T}$ and $\mathbf{g}$ fields impose a strong degree of determinism, but cannot fix position on the manifold, since its structure is too meagre to identify the points and positions that it separates and connects.

That's the problem in a nutshell.

Indeterminism, even if only in respect of manifold points, is surely unwelcome. Yet the models must be distinct if manifold points are. Earman and Norton (1987) and Norton (2011) suggest that we can solve this difficulty by exploiting the absence of discernible differences between models. They identify all of them, justifying this by appeal to what they call Leibniz Equivalence. Thus indeterminism would be avoided since there would be only one model in our example. The suggestion takes us back to the use of Leibniz's *Principle*, together with the static shift, in Euclidean space. The shift argument was examined and found wanting in Chapter 1, §7. Things are quite analogous in the case of the manifold. The manifold's uniform smooth topology replaces the uniform Euclidean metric; the diffeomorphisms replace the rigid motions of the static shift. But with the manifold, again, the shifter must make his "acid test" question explicit: there is no intelligible mapping without specific commitment to distinct manifold points. Does the shifter countenance a model in which distinct points are separated and connected by manifold paths, but in which no point has intrinsic properties at that structural level that could distinguish it (make it discernible)? If so, then Leibniz's Principle is rejected in the shift question, and the diffeomorphically equivalent models are distinct on the shifter's own explicit admission. If not, we can't even begin. A diffeomorphism of the manifold is nothing at all if it is not a mapping of manifold points onto *distinct, separated and connected points* that are, nevertheless, indiscernible one from another. If the shifter is to work in the context of GR (or any other theory that can be formulated in generally covariant fashion) he cannot shirk that commitment.

I conclude that the model building techniques inevitably entail different but indistinguishable models. What differentiates them is that the same tensors are assigned to different but indistinguishable manifold points. Leibniz's Principle cannot justify identifying the models. What Norton calls Leibniz Equivalence will be examined in a later section and argued not to justify identifying models.

10.6.2 A serious embarrassment?

At this point, shouldn't we stop worrying and start living? I shall argue in favour of that advice. What we have to bite here is not a bullet but a piece of cake. We don't *want* to know what isn't determined and we can't make sense of how we could know it. The philosophical interest in Norton's solution lies in whether or not it is a *true* and *metaphysical* solution. Is Leibniz's Principle either sufficient or necessary to avoid indeterminism? I will argue that it is neither. We must and can find a pragmatic way out. Nothing philosophical is relevant.

Not only do we not need to care about the infinity of distinct models. GR provides no way to care about it. Their distinctness is devoid of theoretical consequences as is clear from the general form of model structures. The distinctness of manifold points lies in M, the whole apparatus of prediction lies in the $\mathbf{g}$ and $\mathbf{T}$ fields. They are indeed attached to M, but arbitrarily. In that case, we need no metaphysical principle about the ultimate nature of reality when sound pragmatic sense tells us what to do; when you're working with a theory, don't care about differences that the theory gives you no way to care about. Just identify them and move on. It does not say that that there won't come a theory that exploits these differences in some way. Then you'd have to care. We have no such theory and it is not easy to imagine one.

The same problem crops up with another formulation of GR. Hilbert (1916) is explicitly indebted to Einstein, and published in the same year. Hilbert found a suitable action principle and, by arguments in the calculus of variations, arrived at the fundamental equation by that route. The common problem, then, really lies in this: any law that does not entail a unique metric for spacetime must be stated against a metric-free background. That means that any model will breed a plethora of others, indistinguishable from it in all matter-information and all geometric properties that lie within the scope of the variables in the fundamental equation. Strictly, there are infinitely many distinct, diffeomorphically equivalent models. Neither substantivalists nor realists need retreat from their metaphysical beliefs since the distinct, separated manifold places are unavoidable if we talk of diffeomorphisms of the manifold at all. Still, the identification of these models is a pragmatically commendable step. On the other hand, the received view claims to justify the identification as philosophically correct, not merely as pragmatic. The received view rests on the thesis of Leibniz equivalence. Is that a philosophical thesis?

10.6.3 A philosophical issue?

The prominence of Leibniz's name in the discussion suggests that it is, but Leibniz equivalence is not the same thing as Leibniz's *Principle*. It is only "inspired" by it. The Principle is supposed to be an a priori necessary truth, although it is not analytic unless one explicitly makes it so. It is metaphysical right enough but dubious in its more interesting applications. The Principle does not justify the identification of diffeomorphically equivalent models. Leibniz *equivalence* is a more specific and contingent claim as Norton (2011:§5) states it:

> If two distributions of fields are related by a smooth transformation then they represent the same physical systems.

Also Norton restricts its use. However, the metaphysical *Principle* is not to be understood as limited in its applications. Leibniz would not have tolerated restricting it. That is contrary to the status he accorded it as a thesis in philosophy.

Norton sees the restrictions as follows: "The philosophical problem is to know when two inter-transformable structures do in fact represent the same physical situation ..." He goes on:

> A transformation may ... correspond to no real change in the physical reality represented if
> a... the changes in mathematical structures do not manifest themselves in anything observable; and
> b... the laws of the theory are unable to pick between the two structures related by the transformation.

(Op. cit. §10.3.2 "Philosophical Problem of Gauge Freedoms")

It is not quite clear what the first restriction means. It is obvious that no ordinary observation carried out in either model will reveal that a photon emission event, occurring at one point of its manifold, occurs at a different point in another model's manifold. But consider this example: I snap my left fingers and, a bit later, I snap my right. The events are observable and so is the spacetime path between them. I know that there is a model in which the manifold point of the left snap is occupied by the manifold point of the right snap. The models are linked by a diffeomorphism of the manifold into itself. I can observe that difference in terms of place and date. So I can observe a path between the manifold point in the real world and the point that is its location in the mapped manifold, the merely possible world. What I can't observe are the *terms* of the separation and connection relation if asked to observe them *as featureless*. But the path I observe in the real vivid world is the very same entity as a manifold path that distinguishes the points in the different models. Just so, when I see a particular warthog, I see a particular animal but can't see it as without warthog traits. Further, these relations of smooth connection and separation are diffeomorphic invariants. The *observables* at any conceptual level are just the *invariants* of the level. An observable, a path, does distinguish the models. In the full model, paths permeate perception.

The second restriction is the pragmatic proposal I recommended above. No metaphysical principle validly stands behind it. It is not a truth about GR structures. It is not obviously connected with anything as mathematico-physically weighty as gauge freedom. In our context it arises purely from the arbitrariness in attaching tensors to manifold points. That, in turn, is the product of the very meagre properties ascribed at the manifold level of structure.

Norton (loc. cit.) is not clearly consistent in his discussion of all this. In this same section he writes:

> That the two structures are perfect mathematical images of each other is not sufficient to ensure that they must represent the same physical structures. They certainly *may* ..., but they *may not*. To see this consider a three dimensional physical space with Euclidean properties. The mathematical space hosts many flat two dimensional surfaces, each of which can be transformed perfectly into any other. But to say that these transformations are merely gauge transformations is to collapse the three dimensions of the physical space into two dimensions. Each two dimensional surface in the physical space is a perfect copy of every other one; they are not all the same surface. The transformations between them cannot be gauge transformations.

This is a straightforward case where the *Principle* of the Identity of Indiscernibles demands precisely that collapse. In fact the example readily falls into Leibniz's static shift structure. Suppose a Euclidean world containing just three balls with coplanar centres. Transport all three orthogonally to another plane parallel to the first. Leibniz's argument, employing his principle, tells us that the resulting indiscernible difference means that there is no real change. Nothing has happened! This is just the sort of case where Earman and Norton would have us follow Leibniz in the "acid test." The states of affairs, and thus the planes, are the same. The surfaces are collapsed on one another. Leibniz equivalence also seems fairly clearly to demand it. The Principle clearly does. Yet Norton insists on a distinction.

He does here distinguish mathematical space from physical space but does not show how the difference cuts any ice. The concession about the surfaces in physical space as "perfect copies" is enough to make the case as strong as the static shift case in Euclidean space. But clearly the surfaces are separated and connected by paths – many paths. Unless we outlaw paths as relations that properly discern the states of affairs, the Principle does not apply. Here Norton, unlike Leibniz, is just assuming that spatial relations distinguish but do not identify them. If they are admitted here they should also be admitted in Leibniz's original example and in Earman and Norton's. If they are not admitted, then the Principle demands that flat surfaces in Euclidean 3 dimensional space collapse onto each other. A cascade of collapses must then follow till we end with a single point and disaster.

Norton (op. cit.) also suggests what appears to be a metaphysical problem with manifold substantivalism. It endows the manifold with superfluous "phantom" structure (see second paragraph of the unnumbered introduction to the paper).

One view is that spacetime is a substance, a thing that exists independently of the processes occurring within spacetime. That is spacetime substantivalism. The hole argument seeks to show that this leads to unpalatable conclusions in a large class of spacetime theories. Spacetime substantivalism requires that we ascribe such a surfeit of properties to spacetime that neither observation nor even the laws of the relevants spacetime theory itself can determine which are the correct ones. But there must be some bounds on how rich a repertoire of hidden properties can be ascribed to spacetime.

If my presentation of the argument is correct, most of this introductory paragraph is mistaken. In the realist account of the Hole there is neither surfeit of properties nor any phantom ones ascribed to the manifold. There is no appeal to haecceity. On the contrary! Manifold points have no intrinsic properties at all (unless being 0−magnitude and 0−dimensional count as a properties). They are merely separated and connected by spatiotemporal relations (not even characterised as spatiotemporal at the manifold level of description). In the basic model structure, $< M, \mathbf{g}, \mathbf{T} >$, M's meagre properties are entirely independent of $\mathbf{g}$ and of $\mathbf{T}$ and vice versa. Neither realists nor substantivalists should deny that sense of independence. Realism need not tie itself to the strong view that spacetime could exist without matter. That remains an interesting possibility.

10.7 Einstein on general covariance, 1915 and later

These worries originated in somewhat similar claims made by Einstein (1916). He introduced general covariance to physics informally in §3. It is the most metaphysical section in the paper. He makes the following ontological claims (op. cit. 117):

> We therefore reach this result: In the general theory of relativity, space and time cannot be defined in such a way that differences in the spatial co-ordinates can be directly measured by the unit measuring-rod, or differences in the time co-ordinate by a standard clock ... *The general laws of nature are to be expressed by equations which hold good for all systems of co-ordinates, that is, are co-variant with respect to any substitutions whatever (generally co-variant)* (Original emphasis)... That this requirement of general covariance, which takes away from space and time the last

remnant of physical objectivity, is a natural one will be seen from the following reflexion. All our space-time verifications invariably amount to a determination of space-time coincidences. ... the results of our measurings are nothing but verifications of such meetings of the material points of our measuring instruments with other material points, coincidences of the hands of a clock and points on the clock dial, and observed point-events happening at the same place at the same time. (Emphasis in original).

This account of general covariance is inaccurate: we may not use "*any substitutions whatever*" in coordinates. There are infinitely many discontinuous non-smooth mappings from $\mathbf{R}_4$ into spacetime. Many are not $1-1$. Many do not map number neighbourhoods to spacetime neighbourhoods. As we have seen, the smooth topology of both $\mathbf{R}_4$ and spacetime must be respected in each coordinate chart of any model. Diffeomorphisms are the only transformations allowed. In spacetime, relations of separation and connection are *invariants of the general group of active transformations*. Even if one did think of them as merely a "last remnant of physical objectivity" for space and time, general covariance does not take them away. Without them there is no way to make tensor fields from tensor point assignments. Space, time and spacetime are basic to GR.

The geometrical ideas of GR were new in the physics of 1916 both for readers of the paper and for Einstein himself. To say that he welcomed the conclusion drawn in §3 understates the satisfaction he evidently took in it. Other ontological conclusions that seemed to him to follow from this operationalist thinking was congenial to him well before 1916. Other arguments (op. cit. §2: 112-3) assume that the demands of Mach's Principle are fulfilled in the theory. They aren't. At p. 114, he argues that there is no way to choose between an observer at rest in a local inertial frame and one who accelerates relative to it, to determine which is "really" at rest. The arguments of Chapter 8 show that this is not so. You choose the observer whose worldline is straight. At any moment, she is at rest in a local inertial frame.

Einstein emphasises the role of point coincidences but they are not the immediate invariants of diffeomorphisms. Points are transformed – mapped to other points! Material coincidences merely "tag along." Of course, none is lost or disconnected. However the true invariants are the spatial relations of separation and connection – the topology. Further our observations can't always be of point coincidences, since it is crucial to observe that some coincidences occur elsewhere and elsewhen from others, being smoothly separated and connected by spacetime intervals. Einstein is reported to have said "Time is necessary to stop everything from happening at once."[7] If so, it is interesting that he neglects that role for the manifold in §3. Norton writes of this point coincidence argument (op. cit. §6) "He [Einstein] urges that the physical content of a theory is exhausted by the *catalog* of the spacetime coincidences it licences" (my emphasis). But no theory in physics is a mere catalog – an unstructured *list*.[8] Once again, separation/connection is at risk here. We observe, say, that the absorption of a photon coincides with a reading on the hands of a clock. It is no less crucial that we also observe that the emission of a photon that coincides with another reading on a clock is a distinct spacetime coincidence, the one coincidence being separated from and connected with the other. Model building bottom-up is committed from the outset to the distinctness of smoothly separated and connected manifold points. Without that, no criterion tells us that tensors are being spread over distinct points to make tensor fields. No point has *identifying* properties at the manifold level of structure but they certainly have *distinguishing* relations of separation and connection at that level. The points are all perfectly alike yet separated and connected by paths. So we can begin model building only with an assumption flatly inconsistent with the Principle of the Identity of Indiscernibles. There can be no spreading of fields in spacetime in GR without that smooth topology. It is what stops everything from happening in the same place and time.

Norton goes on (loc. cit.) "Over the years, the hole argument was deemed to be a trivial error by an otherwise insightful Einstein." If the arguments of this chapter are sound, then Einstein's insight did not fail in that judgement but rather in his 1916 remarks on general covariance and point coincidences.

In claims like these, it was hope, rather than hard argument, that fathered Einstein's thoughts. He was never comfortable with realism about space or spacetime. He later thought that, in the absence of matter, GR entails that there is neither space nor time. Yet he knew of, and in his 1916, touched on, matter-free solutions to the equations (op. cit. §15). He stressed the superior importance of free creativity and imagination, regarding them as having been most valuable to him in his intellectual life. That merit was never more brilliantly evident than in his exploitation of spacetime in GR. This book is aimed at defending the view that his superb flights of invention enabled him, and enable us, to think coherently and insightfully about spacetime. It was only on the morning after, so to speak, that he tried to get his

---

[7] A Google search ascribes this to Einstein but gave no source. It has also been said by Wheeler, Hawking and others.

[8] Norton uses "catalog" earlier in a sense importantly different from the one I describe. See (op. cit. §1 "Metrical structure and matter fields") to refer to a list "that specifies the spatial or temporal distance between every pair of events along every curve connecting them."

genie back in the bottle. It is an irony that he was wont to view his own supreme achievement with suspicion and hostility.

10.8 The role of the manifold

An important positive point is salient in this discussion of the active transformation version of general covariance. It shows that three rock bottom spatiotemporal features do not derive from matter or from the field equation. What are fundamental are the primitives of extension (separation and connection), continuity and smoothness. The field equation links the distribution of matter with the metric, the affinity and so on. Each constrains the other. But GR never envisages a way in which matter might generate bare spacetime extension – elsewhere/when-ness – or continuity or smoothness. These ur-properties of space and time can't spring from the natures of what tensor fields are fields *of* – electromagnetism or metric geometry. Spacetime is constrained by matter in GR, but its foundational, most general, properties are not created by it.

Richer features, including the geometric tensors, are conceived as attached to the manifold. They exist in vector spaces conceived as tangent to manifold points. It is arbitrary up to diffeomorphism, which manifold points they are attached to; attachment is constrained only by the topology essential for tensor fields. That puts the manifold structures, weak though they are, beyond deduction from the contents – the tensors – of the theory. The fundamental equation presupposes them in its formulation and they are not consequences of it in any interesting sense. They are fundamental primitives of GR. On any view, at the basic level of smooth, continuous extension, spacetime, in its manifold properties, remains an independent posit of GR. I take this to be a contrary of Einstein's remarks about the last remnant of space and time.

11 Concrete Yet Insubstantial

Chapter 11 draws the threads of the book together and focuses on the conjunction "concrete and insubstantial." It revisits the nature of constructive explanation and shows that that it is not causal explanation. It delineates the ontic type of spacetime in a list of 12 features. The argument ends with a brief but important account of the relation between parts of space, e.g. paths, to space as a whole.

11.1 Introduction

"Spacetime is concrete yet insubstantial" is my signature epigram. I now need to clarify it, especially "insubstantial," in the light of the whole story.

There is a terminological briar patch here and, unlike Brer Rabbit, few of us are likely to feel much at home in it. In threading the maze of the concepts cause, causal power, force, action, free fall, explanation and identity it is easy to get painfully snagged somewhere. In particular, my repudiation of "cause" brings me into amiable conflict (with Hugh Mellor, Chris Mortensen and others) where there is agreement in most essentials. Cause has long been a hotly contested concept. The hope of using the word to convey the same sharp meaning to every reader is faint. Nevertheless I don't need an extended theory of cause.

Despite Einstein's "fancy," his soaring powers of imagination and free invention, he could never deny a deep-seated conviction that there is no way that space, time or spacetime could make sense. They were irrecoverably toxic. I have argued that one needs to understand how something can be both concrete yet immaterial and insubstantial. Perhaps he owed too much to Mach. He tried to turn spacetime into something else – to rebottle the genie. It was a retreat to a superannuated conceptual scheme.

11.2 Cause and causal powers

It is an entrenched belief that time itself is not a cause. The belief that space itself is not a cause is also, but perhaps less deeply, entrenched. I endorse these beliefs. If either space or time were part of *any* causal process then they would be part of *every* causal process and in the same way, formally, in each case. When the metric of spacetime varies from GR model to model and from point to point within any model, it remains true that if spacetime were causally involved in any process in some region of spacetime it would be involved in every causal process there. But the plausibility of these beliefs does not rest on regarding time and space as universal causes that need no mention and go without saying. They are *not* causes. Spacetime merely allows direction and measure to forces and causes. In the main, causes relate events to each other, but also processes and more or less enduring states of things to one another. The presence of a cause temporally precedes its effects. If it is an enduring preceding background state of affairs, as the presence of oxygen is as an INUS condition of fire, then there are or could be, regions of spacetime free of it. So space and time are not causal entities. They are the arenas within which causal processes unfold. They have no causal *powers*.

I take cause to be more than pure regularity and more than truth in all possible worlds that have the same pure regularities as our world has. I take cause to entail more than Humeans admit. I won't embark on saying what more.

The entrenched beliefs are classical. How do they fare in relativistic thought? I argue that they should survive in their apotheosis in spacetime. It is not a cause. I would like to *prove* that these beliefs must survive, but this discussion is partly legislative of my usage and partly argument that the legislation is insightful. I boggle at the idea that spacetime has causal powers. That opens a short and well nigh irresistible route to the conclusion that they are substances, kinds of matter, and then to the backward step, taken by Einstein and others, that spacetime is really an ether. I take "dematerialised ether" to be mere longhand for "spacetime." To conclude that spacetime is, after all, really the gravitational field throws the genius-conceived, unprecedented and potent genie out with the bathwater. But here one does struggle ignorantly on Arnold's darkling plain (and mixes one's metaphors).

I fear that "cause" will always invite "causal powers" as a sequel and that will invite, not pure empirical necessary and sufficient condition in a Humean sense, but substantiality – that spacetime is some sort of stuff. Vice versa, "substance" strongly invites "causal powers." That is exactly what should not be thought if we accept the exposition of the constructive explanation of Chapter 8. The thoughts I am most intent on defending are these: that GR has at its core a new ontology, spacetime; that its ontic type, and its explanatory role, were unprecedented and are still unique[1] in physics. At the very least, one interpretation of GR, in particular its earliest interpretation, makes spacetime its new and principally important agent in a new kind of constructive explanation. Other interpretations of GR may be possible but this is metaphysically the most interesting one, presenting to imagination and invention a formidable but surmountable challenge but also an opportunity for enlarging the power and the scope of disciplined imagination.

Strictly the demand for a non-dynamical explanation of free fall should suffice to make a main intention clear: that spacetime has no dynamical role. But, as I read the subtext of the literature, it is hesitation over the ontology of an insubstantial yet concrete explanans that fogs clear sight of the role of identity in GR's constructive explanation. That fog is dispersible and, I hope, now dispersed. If spacetime has a causal power it could only be its power to tell matter how to move and vice versa. But that single power inevitably suggests a dematerialized ether, as both Newton and Einstein suspected, and as discussed in Chapter 9. That is an unnecessary, obscure and backward step. Spacetime is nothing like an ether.

Causes or their absence "make a difference." Space, time and spacetime might not exist. Then there would have been either nothing at all or some spaceless, timeless world. Barbour (1999) has described a world in which time is an abstraction or fiction. If there were no spacetime, there would be, for instance, no relatively moving rod and no relative contraction. That makes a difference but not a causal difference since the role of identities in constructive explanations is inconsistent with a causal relation.

Explanations are needed for understanding; citing causes is a different enterprise, although the two are related. To find an explanation is not, thereby, to find a cause. Reductive explanations work by proposing identities. In relativity some constructive explanations run differently from others. Examples in SR are particularly simple since the projections up from frame-relative structures to the identical ones in spacetime are linear. I have analysed a number of examples from GR where the causal illusion vanishes in *non*-linear projection up from reference frame description to spacetime description. Their linearity

---

[1] I believe it is absolutely unique in ontic type. In discussion, Roy Sorensen suggested that shadows are also concrete insubstantial things. That is an arresting thought and I am grateful for his proposing it. But I take shadows to consist of differences of illumination over distinct areas or volumes rather than entities. Are shadows necessarily cast on things? I think not: the moon has a shadow whether or not it falls on the earth in an eclipse. When it does, the shadow is not the area or volume shaded but rather the *property* of being less lit than neighbouring surfaces or volumes by some opaque intermediary.

gives examples from SR a transparent simplicity. In these, the constructive nature of spacetime explanation stands out with a special clarity. I conclude this central theme by quickly revisiting of two core SR examples illustrating all the features that make it non-causal.

Spacetime structure explains why a rod moving at some uniform speed relative to an inertial frame is contracted relative to that frame. Minkowski spacetime has explanatory but not causal or temporal priority to those features of the example that it explains. It does not explain why there is a rod or why it is in uniform motion relative to anything. The relative motion does not precede the contraction. Neither motion nor contraction is frame invariant so relative motion is not a tensor and not a cause. The explanation is constructive – a rethinking of a space-and-time feature in spacetime concepts: that was argued in detail in Chapter 5. In space-and-time concepts the matter is imperfectly clear because 1905 SR has an imperfectly defined ontology. In spacetime the ontology is different. The rod is identified with a worldsolid and its uniform motion is identified with the straightness of the solid's timelike edges. The speed of the rod relative to the inertial frame is its non-orthogonal orientation in spacetime to the spacelike hypersurfaces of simultaneity of the frame. The length relative to the frame is identical with the length of those non-orthogonal cross sections. The constructive spacetime explanation exploits these identities and cites no causes. The Minkowskian geometry does not cause the "contraction of the rod." True, if the geometry were different the "contraction "would be different. But the counterfactual rests on a contingent identity. Its truth does not bestow a causal status on spacetime structure.

A similar story may be more briefly told about "the slowing of a moving clock." This is even more clearly constructive if we choose a mechanically minimal Langevin light clock as our example. The relative lengthening of the null (light) paths that "drive" the clock is the relative slowing of it and *is* (an aspect of) its relative motion. The motion does not precede the slowing and so on – it does not cause the slowing.

What is at issue here is which principle forms the foundation of ontology. Is it cause or is it spacetime? I don't know which is the more fruitful and illuminating way to answer this very large question but the one that emerges most strongly from the arguments of this book is spacetime. If, as I have suggested, spacetime embraces within it everything that is real and, conceptually, all that is coherent and intelligible, then it can unify ontology in an illuminating way. That is why I legislate my usage as I do.

I labour these points because they form a principal clarification of my describing spacetime as insubstantial (or immaterial). To abandon them would open the way to ascribing causal powers to spacetime and from there the step to substance is persuasive and short. Even ghosts, if there were any, would have causal powers and so they are substances. Spacetime is no ghost.

11.3 Sophisticated substantivalism and structuralism

I have said little about two views that are salient in the literature. The first is sophisticated substantivalism the second structuralism.

Sophisticated substantivalism is clearly a form of substantivalism not a mere realism. It does not repudiate the Leibniz shifts arguments so it does not stress the crucial point that spacetime relations among things and events are not autonomous. They depend on the geometry of the regions that the things and their path-relations inhabit. It does not reject the spurious distinction between "direct" and path-mediated relations. It rightly rejects the doctrine that points or parts of space have primitive identities. Hoefer (e.g. his (1996) (1998)) has been particularly insistent and clear about their role in substantivalist arguments. In its background lies Leibniz's claim that substantivalism overlooks the fact that the points of space are all exactly alike. But, as argued in the next section, it is not a mistaken belief in primitive identity that mainly misleads us. Finally, the naïve or standard claim is that the metric best represents spacetime: it is also the correct one. That the manifold best represents it is the sophisticated view. Even though it is shared by many physicists it is revisionary and needs justification. That is why Earman and Norton argue for it.

Structuralism as I understand it, holds that in the progress of scientific theories and especially in scientific revolutions, the ontology changes while the abstract structures are largely conserved. Therefore it is the continuing structure that philosophers of science should focus on. That may be so in general but not in relativity theory. Space and time as entities remain largely what they always were when they are unified in spacetime and it is rather the abstract structures that have most changed in GR. My focus and emphasis has been on spacetime as concrete, as pervading perception, especially visual experience. It is not an abstract mathematical object.

11.4 Points and parts of spacetime

The examples of Euclidean and spherical space are significant because their points (parts) lack identities despite their being metric spaces. Reflection on them sheds light mainly on the insubstantiality but also on the concreteness of space and spacetime. Newton wrote:

> ... the parts of space derive their character from their positions, so that if any two could change their positions they would change their character at the same time and each would be converted numerically into the other. The parts of duration and space are only understood to be the same as they really are because of their mutual order and position; nor do they have any hint of individuality apart from that order and position which consequently cannot be altered.[2]

Newton denies identity to the parts of space but affirms their distinctness, citing what I have called their connections with and separations from one another. It is paths that distinguish them.[3]

Newton's main focus was on immobility but he may have felt that this obliged him to seek to describe an ontic type for space as well. So his rejection of space as a substance may also partly rest on the claim that a spatial part is nothing but its order properties yet they do not give them even a *hint* of identity. That is how things stand with any fully metrical homogeneous and isotropic space, not just with the manifold. Spatial parts may have properties that identify them where variable curvatures are available as criteria. But that does not serve to give an identity to parts of Euclidean or spherical space. Metric-preserving mappings are symmetries of S_3 space yet mappings that project points equidistantly along geodesics (i.e. shifts) are not: they change the shapes of things. I conclude that merely rejecting primitive identities fails to erase many of the problems it was aimed at solving Leaving aside properties of regions and properties that derive from the character of regions by limit operations, points have no qualities, especially no sortal ones; they have no intrinsic natures.

Spatial relations may connect any kind of entities so long as the entities are not abstract. They are independent of the properties of their relata so it is not absurd to postulate spatial parts as terms in spatial relations. So spatial parts are just entities that stand in spatial relations. Yet spatial relations can't hold themselves up by their bootstraps. Apart from closed paths they need relata as end points. Something must be a structure-instantiating entity somewhere and somehow for realism to survive. But the parts of space need not play that role. Space itself is real, but it is not *made up* of its parts, including its paths, nor analysable (except notionally) into parts that have any kind of ontic independence. Spatial parts and their relations are, ontologically, supervenient on the structure of space. Spacetime,

---

[2] "De Gravitatione et Aequipondio Fluidorum" (Hall and Hall, 1962:136).

[3] We cannot ascribe an identity to points etc. through a property of haecceity – of being the very the parts they are. That rewrites primitive identity. It is perhaps this that Norton (2011) objects to as superfluous structure that substantivalists ascribe to spacetime. I agree with him on this: there is no such property.

not its parts, is the foundation of spatio-temporal relations. Parts of space are just whereabouts that can be "found" in space; they consist only in relations that belong properly to space itself. (See also Healey (1995) Nerlich (2005).)

That is, space is a real individual but its parts are ersatz individuals. Ersatz because they have no qualities, no intrinsic nature no identities and the only relations they have are to space itself, springing from mere order properties. The order properties are not essential as distinct from accidental. There can indeed be an Aristotelian essence of space – its determinable (not its determinate) metrical structure. Space must have some metric in order to be properly a space, but no particular metric. This is an important aspect of the peculiar insubstantial ontic type that space has, and that time and spacetime have, too.

Here we find an important link between insubstantiality and concreteness. Pure insubstantial extension, pure separation and connection allow us to conceive of distinct things that share all qualities and all non-extensional relations.

11.5 Concluding remarks

I list the main features of the ontic type of spacetime, as I have tried to present it:

Spacetime is:

(i) concrete, not abstract, not a substance

(ii) a particular thing, not a general one

(iii) related to perception in unique, concrete, intimate yet indirect ways

(iv) plays a dominant role in the explanation of physical events and processes

(v) the explanations are constructive, not dynamical but kinematical

(vi) it is not part of, or an element within, the network of causes

(vii) although said to be material in having energy, mass, and to be part of its own source, its characteristics remain geometrical, insubstantial and concrete

(viii) it is everywhere and everywhen

(ix) it is *analysable* into parts but not *composed* of ontologically independent or distinct parts

(x) it is not analysable by means of spatial relations among observables

(xi) it contains, relates, orientates and partially measures forces and causes

(xii) the fundamental equation of GR is one of mutual constraint but not mutual causality.

Spacetime has a central role in metaphysics for a number of reasons. It is plausible that every real entity is spatiotemporally related to every other so that spacetime gives a unity to our ontological scheme It may provide a plausible and clear conceptual boundary for all real entities. It may also provide a plausible and clear conceptual boundary round all the things that can be made intelligible. It is not my main aim to defend that view of reality, although it has somewhat guided me. It is an important and interesting one. However, it cannot really work until space, time and spacetime are themselves satisfactorily analysed, and until this criterion of the real is no longer itself disputed and commonly regarded as metaphysically suspect. To some extent that anxiety has always been allayed by the hope that some form of relationism would succeed in dismissing the difficulty. But that cannot succeed if it is true that spatial relations among observables themselves inextricably involve such things as paths and regions. The book began by making out a case that spatial relations are dependent on regions and thus on space-like entities.

If my arguments are successful then I believe a favourable view of them will depend on the degree of their directness and simplicity. My aim throughout has been to make them so.

Last and by no means least, to justify spacetime as a sound and limpid concept would take a significant step in the direction of freeing and enlarging intellectual imagination. Einstein exploited a prodigious flair for fancy and invention, yet if the arguments in this book are sound he was bound by a false conviction that spacetime just couldn't make sense. To have freed himself from that that would have been for him to have recognised his genie for what it really is. I have not tried to solve the deep problem what the bounds of sense and imagination are. But it will be of interest if I have shown that the invention of spacetime did not transgress them.

One shouldn't, and I don't, offer these arguments as closing the issues raised here. My best hope is that they will add something of value to our combined enquiries. In philosophy, but not only there, every apparently final answer merely awaits a better question.

REFERENCES

Alexander, H. (ed.) (1956) *The Leibniz-Clarke Correspondence.* Manchester, Manchester University Press.

Anderson, J. L. (1967) *Principles of Relativity Physics.* NY. Academic Press.

Arthur, R. (2007) "Time, Inertia and the Relativity Principle:" (archived in the PhilSci archives at Pittsburgh); from the Symposium on Time and Relativity (Minneapolis, MN, 25-27 October, 2007).

Bacry, H. and Levy-leBlond, J-M. (1968) "Possible Kinematics." *Journal of Mathematical Physics* **9**: 1605 - 1614.

Baez, J. (2002) "Noether's Theorem in a Nutshell" math.ucr.edu/home/baez/noether.html

Balashov, Y. and Janssen, M. (2003) "Presentism and Relativity" *British Journal for the Philosophy of Science* **54**: 327-46.

Barbour, J. (1999) *The End of Time.* London, Weidenfeld and Nicholson.

Bekenstein, J. (2004) "Relativistic Gravitation Theory for the MOND Paradigm" http://arxiv.org/abs/astro-ph/0403694.

Bell J. S. (1987) "How to teach special relativity" in *Speakable and Unspeakable in Quantum Mechanics: papers on quantum philosophy.* Cambridge CUP.

Bigelow, J. (1996) "Presentism and Properties." *Philosophical Perspectives* **10**:35-52.

Bigelow, J. and R. Pargetter (1989) "Vectors and Change." *British Journal of the Philosophy of Science* **40**: 289-306.

Black, M. (1952) "Identity of Indiscernables." *Mind* **61**: 153-164.

Bourne, C. (2006) *A future for Presentism.* Oxford. OUP.

Bricker, P. (1993) "The Fabric of Space: Intrinsic vs. Extrinsic Relations": 1-37. *Midwest Studies in Philosophy* XV11, P.A. French, T.E. Uehling and H. K. Wettstein, (eds). Notre Dame, IN: Notre Dame UP.

Brighouse, C. (1994) "Spacetime and Holes." *Philosophy of Science Association* **1**: 117-125.

Broad, C. D. (1938) *An Examination of McTaggart's Philosophy.* Cambridge CUP.

Brockman, J. (1995) *The Third Culture: beyond the scientific revolution.* NY Simon and Schuster.

Brown, H.R. and Pooley, O. (2001) "The origin of the spacetime metric: Bell's 'Lorentzian pedagogy' and its significance in general relativity" in C. Callender and N. Huggett (eds.) *Physics meets Philosophy at the Planck Scale: contemporary theories in quantum gravity.* Cambridge, CUP: 256-72.

Brown, H. R. (2005) *Physical Relativity.* Oxford, Clarendon

Brown, H.R. and Pooley, O. (2006) "Minkowski spacetime: a glorious nonentity" in D. Dieks (ed) The ontology of spacetime. Elsevier 67-89.

Butterfield, J. (1984) "Seeing the Present." *Mind* **93**: 161-176.

Butterfield, J. (1987) "Substantivalism and Determinism." *International Studies in the Philosophy of Science* **2**(1): 10 - 32.

Butterfield, J. (1988) "Albert Einstein Meets David Lewis." *Biennial Meeting of the Philosophy of Science Association*, Philosophy of Science Association.

Butterfield, J. (1989) "The Hole Truth." *British Journal of the Philosophy of Science* **40**: 1-28.

Cacciatori, S., Gorini, V., Kamenshchik A. (2008) "Special Relativity in the 21st Century" *Annalen der Physik.* July: 1-40.

Carroll, L. (1871) *Through the Looking Glass and What Alice Found There.* London, McMillan: 109

Chalmers, A. (1993) "Galilean Relativity and Galileo's Relativity." *Correspondence, Invariance and Heuristics: Essays in Honour of Heinz Post.* Ed. S. French and H. Kamminga. Dortrecht, Kluwer Academic Publishers: 189-205.

Clifford, W. K. (1995) *The commonsense of the exact sciences*. NY Dover.

Clifton, T. and Ferreira, P. (2009) "Does Dark Energy Really Exist?" *Scientific American* April 2009

Corry, L. (1997) "Hermann Minkowski and the Postulate of Relativity." *Archive for the History of the Exact Sciences* **51**: 281-314

Corry, L. (2004) David Hilbert and the Axiomatization of Physics, 1898-1918: *From "Grundlagen der Geometrie" to "Grundlagen der Physik"* Dordrecht: Kluwer.

Cox, B. and Forshaw, J. (2009) *Why does $E = mc^2$* Camb. Da capo Press

Craig, W. L. (2001) *Time and the Metaphysics of Reality*. Dordrecht: Kluwer.

Crisp (2003) "Presentism" In Loux, M. and Zimmerman, D. *Oxford Handbook of Metaphysics* Oxford OUP.

Dainton, B. (2006) *Stream of Consciousness*. London Routledge.

Dainton, B. (2010) *Time and Space*. Durham, Acumen.

Davies, P. and Brown J. R. (1986) *The Ghost in the Atom: A Discussion of the Mysteries of Quantum Physics*. Cambridge: Cambridge University Press.

Davies, P. et al. (2006) In "The Anthropic Principle." *The Science Show, Australian Broadcasting Commission*, Radio National

DiSalle, R. (1994) "On Dynamics, Indiscernibility, and Spacetime Ontology." *British Journal of the Philosophy of Science* **45**: 265-287.

DiSalle, R. (1995) "Spacetime Theory as Physical Geometry." *Erkenntnis* **12**: 317-337.

DiSalle, R. (2006) *Understanding Space-Time: the Philosophical Development of Physics from Newton to Einstein*. Cambridge, CUP.

Dyke, H. (2007) "Tenseless/Non-Modal Truthmakers for Tensed/Modal Truths," *Logique et Analyse* **199**: 269-87.

Dyson, F. (1972) "Missed Opportunities." *Bulletin of the American Mathematical Society* **78**: 635-672.

Earman, J. and Friedman, M. (1973) "The meaning and status of Newton's Law of Inertia and the Nature of Gravitational Forces." *Philosophy of Science* **40**: 329-359.

Earman, J. (1973) "Covariance, Invariance, and the Equivalence of Frames." *Foundations of Physics* **4**(2): 267 - 289.

Earman, J. (1986) *A Primer on Determinism*. Dordrecht, Reidel.

Earman, J. and J. D. Norton. (1987) "What Price Spacetime Substantivalism? The Hole Story." *British Journal of the Philosophy of Science* **38**: 521.

Earman, J. (1989) *World Enough and Space-Time: absolute versus relational theories of space and time*. MIT Press, Cambridge.

Earman, J. (2006) "The Implications of General Covariance for the Ontology and Ideology of Spacetime." In *The Ontology of Spacetime*. D. Dieks (ed,), Elsevier: 3-23.

Eddington, A. (1928) *The Nature of the Physical World*. Cambridge, CUP.

Ehlers, J., Pirani, F. and Schild, A. (1972) "The Geometry of Free Fall and Light Propagation." In O'Raifeartaigh, L. *General Relativity* Oxford OUP.

Einstein, A. (1905) "On the electrodynamics of moving bodies." In Lorentz et al. (1923)

Einstein, A. and Fokker, A. (1914) "Die Nordströmsche gravitationstheorie vom Standpunkt des absoluten Differentialkalküls" *Annalen der Physik* **44**: 321-8.

Einstein, A. (1916) "The Foundations of the General Theory of Relativity." In Lorentz et al. (1923)

Einstein, A. (1918) "Prinzipielles zur allgemeinen Relativitätstheorie "Annalen der Physik. 55: 241-4.

Einstein A. (1919) "What is the Theory of Relativity?" *In Ideas and Opinions* ed. C. Selig. NY Crown Publishers Inc. original in The London Times, November 28, 1919

Einstein A. (1953) *The Meaning of Relativity*. Princeton, Princeton University Press.

Einstein, A. (1954) *Relativity: the Special and General Theory: a Popular Exposition*. 15th edition. London. Methuen.

Einstein, A. and Infeld, L. (1961) *The Evolution of Physics: from early concepts to relativity and quanta*. NY, Touchstone Books.

Einstein, A. (1983) *Sidelights on Relativity*. NY Dover.

Everett, H. III. (1953) "Relative State Formulation of Quantum Mechanics." *Reviews of Modern Physics* **29**: 454-462.

Feigenbaum, M. (2008) "The Theory of Relativity – Galileo's Child." http://arxiv.org/abs/0806.1234 (2008).

Fernflores, F. (2011) "Bell's Spaceships Problem and the Foundations of Special Relativity." *International Studies in the Philosophy of Science*, **25**:4, 351-370.

Feynman, R. (1965) *The Feynman Lectures on Physics*. Palo Alto, Addison-Wesley.

Feynman, R. (1985) *QED: The Strange Theory of Light and Matter*. Princeton, University Press.

Fitzgerald, G. (1889) "The Ether and the Earth's Atmosphere" *Science* **13**:390.

Forrest, P. (2004) "The real but Dead Past." *Analysis* **64**:358-62.

Friedman, M. (1983) *Foundations of Space-time Theories: Relativistic Physics and Philosophy of Science*. Princeton, Princeton University Press.

Gamow, G. (1957) *Mr Thompkins in Wonderland: or, stories of c, G and h*. Cambridge, Cambridge University Press.

Geroch, R. P. (1978) *General Relativity from A to B*. Chicago, Chicago University Press.

Gödel, K. (1949) "A Remark About the Relationship Between Relativity Theory and Idealistic Philosophy" in Schilpp (1949)

Graves, J. (1971) *The Conceptual Foundations of Contemporary Relativity Theory*. Cambridge Mass, MIT Press

Grünbaum, A. (1973) *Philosophical Problems of Space and Time*. In *Boston Studies in the Philosophy of Science* vol 12 Dordrecht: Reidel.

Haldane, J.B.S. (1930) Possible Worlds and other Essays. London, Chatto and Windus.

Hall, A. R. and Hall, M. B: (1962) *Unpublished Scientific Papers of Isaac Newton*, (eds. and trans.) Cambridge, Cambridge University Press.

Hawking, S. and Ellis, G. (1973) *The Large Scale Structure of Space-Time*. Cambridge, CUP.

Healey, R. (1995) "Substance, Modality and Spacetime." *Erkenntnis* **42**, 287-316.

Helmholtz, H von (1960) "On the Origin and Significance of Geometrical Axioms." In James R. Newman (ed) *The World of Mathematics*. London, Allen and Unwin: 647-68.

Hilbert, D. (1916) "Die Grundlagen der Physik (Erste Mitteilung)." *Nachtriten von der Königlichen Gesellschaft der Wissen schaften zu Göttingen. Mathematisch-Physilalische Klasse* 39-407.

Hoefer, C. (1996) "The Metaphysics of Space-Time Substantivalism." *The Journal of Philosophy* **XCIII**: 5-27.

Hoefer, C. (1998) "Absolute versus Relational Spacetime: For Better or Worse the debate Goes On." *British Journal of the Philosophy of Science* **49**: 451-467.

Hoefer, C. (2000) "Energy Conservation in GTR" *Studies in History and Philosophy of Modern Physics* **31**: 187-199.

Hooker, C. (1971) "Relational Doctrines of Space and Time" *British Journal for the Philosophy of Science* **22**: 97-130.

Hudson, H. (2005) *The metaphysics of Hyperspace*. Oxford, Clarendon Press.

Huggett, N (2000) "Reflections on Parity Nonconservation" *Philosophy of Science* **67**: 219-41.

Huggett, N. and Hoefer, C. (2006) "Absolute and Relational Theories of Space and Motion." *Stanford Encyclopedia of Philosophy*.

Ignatowski, V. S. (1910) "Einige allgemeine Berkummengen zum Relativitäsprinzip" *Phys. Ztschr.* **11**: 972-6.

Janssen, M. (1995) "A comparison between Lorentz's ether theory and Einstein's special theory of relativity in the light of the experiments of Trouton and Noble," *Ph.D. Thesis*, University of Pittsburgh.

Janssen, M. (2009) "Drawing the line between kinematics and dynamics in special relativity." *Studies in History and Philosophy of Science Part B: Studies in History and Philosophy of Modern Physics* **40**: 26-52.

Kant, I. (2007) *Critique of Pure Reason*. Trans. Norman, K.S. and Barham,G Basingstoke, Palgrave Macmillan.

Kretschmann, E. (1917) "Ueber des Physikalischen Sinn Relativitätspostuaten." *Annalen der Physik* **53**: 575-614.

Ladyman, J. (2009) "Structural Realism." *The Stanford Encyclopedia of Philosophy* (Summer 2009 Edition), Edward N. Zalta (ed.).

Lange, M. (Forthcoming 2013) "How to explain the Lorentz transformations." In Mumford, S. and Tugby, M. (eds.) *The Metaphysics of Science.* Mind Assoc, Occasional Series.

Lange, M. (2002) *An Introduction to the Philosophy of Physics: Locality, Fields, Energy and Mass.* Oxford, Blackwell.

Lehmkuhl, D. (2011) "Mass-energy-momentum: Only There Because of Spacetime?" *British Journal of Philosophy of Science* **62**: 453-488.

Leibniz, G. W. (1898) *The Monadology and Other Philosophical Writings.* Trans. R. Latta. Oxford Clarendon Press.

Lorentz H. A. (1892) "The relative ion of the Earth and the Aether." *Zittingsverlag Akad. V. Wet* **1**: 74-7

Lorentz, H. A. (1904) "Electromagnetic phenomena in a system moving with any velocity less than that of light." In Lorentz et al. (1923).

Lorentz et al. (1923) *The Principle of Relativity: a collection of original memoirs on the special and general theory of relativity.* NewYork: Dover.

Lucas, J. R. (1984) *Space Time and Causality: an Essay in Natural Philosophy.* Oxford, Clarendon Press.

Lucas, J. and Hodgson, P. (1990). *Spacetime and Electromagnetism.* Oxford, Clarendon.

Lyle, S. (2010a) *Self-force and inertia: Old light on New Ideas.* Berlin Springer.

Lyle, S. (2010b) "Rigidity and the Ruler Hypothesis" in V. Petkov (ed.) *Space, Time and Spacetime: Physical and Philosophical Implications of Minkowski's Unification of Space and Time.* Springer: 61-106.

Malament, D. (1986) "Time Travel in the Gödel Universe" *PSA 1984, vol. 2*; (Proceedings of the Philosophy of Science Association meetings, Chicago, 1984)

Maudlin, T. (1990) "Substances and Space-Time: What Aristotle Would Have Said to Einstein." *Studies in History and Philosophy of Science* **21**: 531-61.

Maudlin, T. (2007) *The Metaphysics within Physics.* Oxford, OUP.

McCall, S. (1994) *A Model of the Universe: Space-Time, probability and decision.* Oxford, OUP.

McCall, S. and Lowe, E. J. (2003) "3D/4D Equivalence, the Twins Paradox and Absolute Time." *Analysis* **63**: 114-23.

Mellor, D. H. (1998) *Real Time II.* London, Routledge.

Mermin, N. D. (1984) "Relativity Without Light." *American Journal of Physics* **52**: 119 - 24.

Mills, R. (1989) "Gauge Fields." *American Journal of Physics* **57**: 493 - 507.

Minkowski, H. (1908) "Space and Time." In Lorentz H. A. et al. *The Principle of Relativity*, Dover, New York (1923).

Minkowski, H. (1915) "Das Relativitätsprinzip." *Journal of the Deutsche Mathematiker-Vereinigung (DMV).* **24**:372-382.

Misner, C., Thorne, K. and Wheeler, J. (1973) *Gravitation.* New York Freeman.

Nerlich, G. (1991) "How Euclidean Geometry Has Misled Metaphysics." *Journal of Philosophy*, **88**:169-89

Nerlich, G (1994) *The Shape of Space* 2nd ed. [(1976) 1st edn.] Cambridge, New York, Cambridge University Press.

Nerlich, G. (1998) "Time as Spacetime." *In Questions of Time and Tense.* R. Le Poidevin. Oxford, Clarendon Press: 119-134.

Nerlich, G. (2004) "How the Twins Do It: Special Relativity and the Clock Paradox." *Analysis* **64**: 219-230.

Nerlich, G (2005) "Can the Parts of Space Move? On paragraph 6 of Newton's Scholium." *Erkenntnis* **62**: 119-135.

Nerlich, G. (2010) "Why spacetime is not a hidden cause: a realist story" in V. Petkov (ed.) Space, Time and Spacetime Physical and Philosophical Implications of Minkowski's Unification of Space and Time: Springer: 177-188.

Newton, I. (1999) *The Principia: Mathematical Principles of Natural Philosophy* (Trans. I. B. Cohen and A. Whitman) Berkeley, University of California Press. p 416

Noether, E. (1918). "Invariante variationsprobleme," *Göttinger Nachrichten, Math.-phys. Kl.*, 235-57.

Norton, J. D. (1985) "What was Einstein's Principle of Equivalence?" *Studies in History and Philosophy of Science* **16**: 203-246.

Norton, J. D. (1987) "Einstein, the Hole Argument and the Reality of Space." In *Measurement, Realism and Objectivity*. J. Forge(ed.) D. Reidel Publishing Company: 153-188.

Norton, J. D. (1988) "The Hole Argument." Philosophy of Science Association 2: 56-63.

Norton, J. (1992) "Philosophy of Space and Time" in Salmon, M. (ed) *Introduction to the Philosophy of Science*. Eglewood Cliffs, Prentice-Hall. Chapter 5.

Norton, J. D. (1993) "General Covariance and the Foundations of General Relativity," Sect 2, *Reports on Progress in Physics* **56**:791-858.

Norton, J. D. (2008) "Why Constructive Relativity Fails." *British Journal for the Philosophy of Science* **59**: 821-34.

Norton, J. D. (2010) "Einstein's Special Theory of Relativity and the Problems in the Electrodynamics of Moving Bodies that Led him to it." in *Cambridge Companion to Einstein*, M. Janssen and C. Lehner, eds., Cambridge University Press.

Norton, J. D., (2011) "The Hole Argument." *The Stanford Encyclopedia of Philosophy* (Fall 2011 Edition), Edward N. Zalta (ed.).

Patterson, E, M. (1969) *Topology*. Edinburgh, Oliver and Boyd.

Paul, L. A. (2010) "Temporal Experience." *Journal of philosophy* **CVII**: 333-59.

Pauli, W. (1958) *The Theory of Relativity*. London, Pergamon.

Petkov, V. (2009) *Relativity and the Nature of Spacetime* 2nd edn. NY Springer.

Petkov, V. ed. (2010) *Space, Time, and Spacetime – Physical and Philosophical Implications of Minkowski's Unification of Space and Time*, Berlin, Heidelberg, New York: Springer.

Petkov, V. (2012) "Can Gravity be Quantised?" Appendix C of his *Inertia and Gravitation: From Aristotle's Natural Motion to Geodesic Worldlines in Curved Spacetime*. Minkowski Institute Press.

Pooley, O. (2003) "Handedness, Parity and the Reality of Space." In Brading, K. and Castellani, E. (eds.) *Symmetries in Physics: Philosophical Reflections*. Cambridge CUP.

Prior, A. (1968) *Papers on Time and Tense.* Oxford, Clarendon.

Prior, A. (1970) "The notion of the present." *Studium Generale* **23**: 245-48.

Price, H. (1996) *Time's Arrow and Archimedes' point: New Directions for the Physics of Time.* Oxford, OUP.

Quine, W. V. (1960) *Word and Object.* Cambridge, MIT Press.

Rea, M. (2003) "Four dimensionalism" In Loux, M. and Zimmerman, D. *Oxford Handbook of Metaphysics.* Oxford OUP.

Rees, M. (1999) *Just 6 numbers: the deep forces that shape the universe.* London, Weidenfeld and Nicholson.

Resnick, R. (1968) *Introduction to Special Relativity.* N.Y.: Wiley.

Rindler, W. (1977) *Essential Relativity: Special, General and Cosmological*, New York, Van Nostrand Reinhold Co

Rindler, W. (2001) *Relativity: Special, General and Cosmological*, Oxford, OUP

Rovelli. C. (1997) "Halfway through the woods: contemporary research on space and time." In John Earman and John D. Norton, (eds.), *The Cosmos of Science*, 180-223. University of Pittsburgh Press, Pittsburgh

Rynasiewicz, R. (1994) "The Lessons of the Hole Argument." *British Journal for the Philosophy of Science*, **45**: 4-07-36.

Rynasiewicz, R. (1996) Absolute vs. Relational Spacetime: An Outmoded Debate?, *Journal of Philosophy* **93**: 279-306

Savitt, S (1996) "The Direction of Time" *British Journal for the Philosophy of Science*, **47**: 347-70

Savitt, S. F. (2006) "Being and Becoming in Modern Physics." *Stanford Encyclopedia of Philosophy* http://,plato.stanford.edu/entries/spacetime-bebecome/ (accessed March 2010)

Schilpp, P. A. (ed.) (1949) *Albert Einstein: Philosopher-Scientist* NY Harper.

Schrödinger, E. (1950) *Space-Time Sructure.* Cambridge. CUP

Schutz, B. (1985) *A first course in general relativity*, Cambridge, CUP

Sen, A. (1994) "How Galileo Could Have Derived the Special Theory of Relativity" *American Journal of Physics* **62** issue 2 (Feb 1994) p. 157 (with diagrams)

Sklar, l. (1974a) "Incongruent counterparts, intrinsic features and the substantiviality of space" *Journal of philosophy*, **71**: 277-90.

Sklar, L. (1974b) *Space, Time and Spacetime*. Berkeley and Los Angeles, University of California Press.

Stein, H. (1968) "On Einstein-Minkowski Space-Time." *The Journal of Philosophy*, **65**(1): 5-23.

'tHooft, G. (1980) "Gauge Theories of Forces Between Elementary Particles." *Scientific American*, **242**(6): 104-16.

Taylor, E. F. and Wheeler, J. A. (1992) *Spacetime Physics* 2nd edn. NY Freeman.

Taylor, E. F. and Wheeler J. A. (2000) *Exploring Black Holes: an introduction to general relativity*. San Francisco, Addison Wesley Longman.

Time Magazine (2000) Front cover 1st issue. Chicago Time Inc.

Tooley, M. (1997) *Time, Tense and Causation*. Oxford: Oxford University Press.

Torretti, R. (1978) *Philosophy of Geometry from Riemann to Poincaré*. Dordrecht-Holnd, Boston, D.Reidel Pub Co

Torretti, R. (1983) *Relativity and Geometry*. Oxford, Pergamon 79-80

Torretti, R. (1999) *The Philosophy of Physics*. Cambridge, Cambridge University Press.

Van Cleve, J. and Frederick, R. (eds.) (1991) *The Philosophy of Left and Right*. Dordrecht, Kluwer.

Wald, R. (1984) *General Relativity. Chicago*, Chicago University Press.

Walter, S. (2010) "Minkowski's Modern World" in V. Petkov, ed, *Minkowski Spacetime: A Hundred Years Later*, Springer, 2010, pp. 43-61. A corrected version is available from Walter's website.

Walter, S. (1999) "Minkowski, Mathematicians and the Mathematical Theory of Relativity" in Goenner, H. et al (eds) *The Expanding Worlds of General Relativity* (Einstein Studies 7) : 45-86 Boston, Birkhauser. 1999

Walter, S. (2008) "Hermann Minkowski's Approach to Physics" *Mathematische Semesterberichte*, **55**(2): 213-235

Walter, S. (2011) "Figures of Light in the Early History of Relativity (1905 - 1914)" to appear in Rowe, D. (ed) *Einstein Studies 12* Boston, Birkhauser: preprint at www.univ-nancy2.fr/DepPhilo/walter

Weyl, H. (1952) *Symmetry*. Princeton, Princeton University Press.

Wheeler, J. A. (1998) *Black Holes, Geons and Quantum Foam*. London. W. W. Noeton

Wheeler, J. A. (1990) *A Journey into Gravity and Spacetime*. NY Scientific American Library

Whittaker, E.T. (1910) *A History of the Theories of Aether and Electricity: Vol 2 The Modern Theories 1900-1926. Chapter II: The Relativity Theory of Poincar and Lorentz*, Nelson, London.

INDEX

'tHooft, G., 57
3D language, 116
3D/4D equivalence, 114
4-momentum, 150
4-tensor, 91, 126
4-vector, 73, 96
4D world, 102

Absolute rest, 13, 117–118
Acceleration
    Gentle, 84–96
    Rindler, 95–96
Acid test, 12–25, 180, 183
Aladdin, 2, 5
Anderson, J. L., 152
Arnold, M., 6
Arthur, R., 56
Augustine, 109

Backfire, 16
Baez, J., 152
Balashov, Y., 75, 78, 83
Barbour, J., 191
Bell, J. S., 76, 78, 80, 83, 93
Bigelow, J., 105–107, 111
Black, M., 45
Bourne, C., 104
Bricker, P., 37
Broad, C. D., 100
Brown, H., 56, 57, 75, 78, 88, 123, 134, 135, 161
Butterfield, J., 112, 177

Cartesian coordinates, 167, 169
    extendible, 171
Cauchy surface, 177

Cause
    defined, 129
    Not identity, 190–193
CERN, 86
Chalmers, A., 125
Clarke, S., 11
Clifford, W. K., 5
Clifton, T., 54
Clock Paradox, 5, 100, 113
cone
    light, 73
    null, 73
conservation principles, 68
constructive spacetime explanation, 192
contraction
    cause of, 80, 82
    Fitzgerald, 79
    Lorentz, 79
conventionalism, 48
Copernican Revolution, 40
Corry, L., 49
Coulomb field, 84
covariance
    general, 166
    special, 166
Craig, W. L., 117
creativity of intellect, 5

Dainton, B., 37, 104, 112, 146
Davies, P., 57, 83
de Sitter spacetime, 54, 160
Dennett, D., 14
Descartes, R., 7
determinism, 90, 176
diffeomorphism, 174, 175, 187

differential manifold, 171
DiSalle, R., 7, 123
disciplined imagination, 5
divergence of **T**, 160–162
Doppler red shift, 42
dualism, 18, 24
Dyke, H., 109, 112
Dyson, F., 48

Earman, J., 12, 15, 20, 25, 28, 126, 146, 151, 165, 177, 193
East on Earth, 15
Eddington, A. S., 4, 77
Ehlers, J., 134
Einstein
    Imagination, 4
Einstein tensor, 161
Einstein's relativity theories, 1
Einstein, A., 3, 5, 48, 76, 83, 90, 126, 132, 139, 142, 165
electromagnetism, 57
energy
    from stars, 156–157
    kinetic, 64, 69, 71, 77
    localised, 154
    potential, 65
    total, 65
    two strands, 153
energy-momentum, 63
energy-momentum vector, 69, 73
epistemology, 8
Equivalence Principle, 163
Euclidean geometry, 5
Everett diagram, 101, 102, 111
Everett, H. III., 111
explanans, 77
explanation
    constructive, 77, 90
    existential, 124
    geometrical, 124
Extremal aging, 127

Faraday, M., 4
Feigenbaum, M., 56

Fernflores, F., 96
Ferreira, P., 54
Feynman, R., 128, 152
Fields
    Spreading of, 179–186
Fitzgerald contraction, 79, 82, 85, 87, 117
Fitzgerald, G. F., 79
Fitzgerald-Lorentz contraction, 87
Flow of Time, 100
flux
    spacelike, 150
    timelike, 150
force, 67
    tidal, 132
Forrest, P., 104
Free mobility, 20
Friedman, M., 126

Gödel, K., 177
Galilean relativity, 55
Galileo, 125
Galileo 4-trajectories , 129
Gamow, G., 57, 94
Gauss's Law, 161
Gauss, C. F., 36
general covariance, 174, 189
General Relativity, 1, 4
generally covariant, 173
genie
    dark side, 5
geometry
    Euclidean, 20, 55, 166
    hyperbolic, 68
    of constant curvature, 20
    projective, 20
Geroch, R., 151
Graves, J., 147
gravitational field, 127, 130, 143
gravitational force
    tidal, 130
gravitational red shift, 42
gravitational self-energy, 159
gravitational waves, 156–157
graviton, 56

gravity, 124, 130, 132, 153
    as a force, 2, 127
        causal in reference frame, 129
        not a cause, 124
Greek Mythology, 105
gutters, 135

Healey, R., 195
Helmholtz, H., 20
Hilbert, D., 49
Hoefer, C., 146, 159, 161, 162, 193
Hole Argument, 12, 23, 165, 174, 176
Hole argument
    Surfeit of properties, 184

ideology, 77, 93
Ignatowski, V. S., 56, 69
Indexicals, 9
indexicals, 109–110
initial value problems, 151
intellectual imagination, 6
interval
    spacelike, 96
intuition pumps, 14
isotropy
    kinematic, 55

Janssen, M., 75, 83, 92

Kant, I., 29, 40
kinematical motion
    pure, 79
Kretschmann, E., 171

Ladyman, J., 177
Lange, M., 75, 93
Langevin clock, 117, 192
Larmor, J., 81, 83
Last remnant, 185
last remnant of physical objectivity, 2, 166, 169, 185
Lehmkuhl, D., 58, 151, 159
Leibniz
    Equivalence vs Principle(, 176
    Equivalence vs Principle), 183

Leibniz Equivalence, 19, 20, 23, 24, 46, 176
Leibniz shifts, 11, 25, 32, 175
Leibniz, G. W., 11, 13, 20, 23, 38
Leibnizian monad, 39
Lie groups, 48
Lorentz contraction, 79, 86, 94, 96
Lorentz coordinates, 170
Lorentz group, 54
Lorentz invariant, 85
Lorentz transformations, 92
Lorentz, H. A., 56, 79, 83
Lorentzian ether, 145
Lorentzian pedagogy, 78, 81, 83, 89
Lowe, E. J., 100, 113
Lyle, S., 96

Mach's Principle, 185
Mach, E., 6
Malament, D., 177
manifold
    differentiable, 173
mass
    "lost", 71
    proper, 63
    relativistic, 63
materialism, 7
Maudlin, T., 100
Maxwell's electrodynamics, 55, 88
Maxwell's electromagnetism, 5
Maxwell's equations, 81, 83, 91, 93, 117
Maxwell, J. C., 4
Maxwell-Lorentz theory, 85
McCall, S., 100, 101, 107, 110, 111, 113
Mellor, D. H., 100, 109
Mellor, H., 189
Mercury
    advance of the perihelion, 4
Mermin, N. D., 56
metalinguistic sense, 109
metaphysics, 1, 6, 35, 40, 58
metric

Euclidean, 51
intrinsic, 36
Lorentz, 48
metric field, 7
metric spaces, 12
Minkowski diagrams, 9
Minkowski geometry, 115
Minkowski spacetime, 54, 56, 61, 78, 91, 97, 127, 129, 146, 151, 158, 167, 173, 192
Minkowski's ontology, 97
Minkowski, H., 5, 47, 49, 54, 57, 63, 69, 90, 132
Misfire, 16
Misner, C., 96
Modest Realism, 43
Monadology, 38
Mortensen, C., 189
mystery of mysteries, 136
mystification, 80–83

Nerlich, G., 17, 32, 112, 123, 135, 195
Newton's first law, 58
    and cause, 125
    and trajectories, 126
Newton's third law, 141
    reciprocal action, 144–145
Newton, I., 6, 11, 33, 125
Newtonian physics, 151
Noether's theorems, 152
Norton, J., 11, 12, 15, 20, 23, 25, 49, 75, 88, 175, 176, 193

O'Hair, G., 136
Observable changes, 182
ontic type, 1, 6, 195
ontic types, 44
    spooky, 8
ontology, 1, 6, 40, 45, 58, 77, 93
    indeterminate in 1905 SR, 89
    reductive, 93
    unify in an illuminating way, 192

path, 33

extremal, 36
orientation in spacetime, 34
spacetime, 33
path realism, 40
Paul, L. A., 112
Pauli, W., 56
Petkov, V., 54, 89, 91, 148
photon, 72
photon vector, 73
physics, 58
Pirani, F., 134
Plato, 100, 109
Poincaré group, 54
Poincaré's conventionalism, 49
Poincaré, H., 13, 49, 81
Point coincidences, 185–186
Pooley, O., 75, 78, 88, 135
present chauvinism, 112
presentism, 100, 104, 105
presentness, 104
Principia, 125
Principle of Equivalence, 127, 142, 161
Principle of Sufficient Reason, 13
Principle of the Diversity of Discernibles, 20
Principle of the Identity of Indiscernibles, 13, 17, 20, 22
Prior, A., 100, 106, 107, 111, 117
pseudo-Riemannian metric, 174
Pythagoras's theorem, 50

quantum gravity, 10
quantum theory, 2
Quigley, P., 37

radiation, 72
radical break, 80–83
rapidity, 67
Rash Realism, 44
realism, 18, 35, 41, 43, 124
    best objection to, 123–124
    minimal
        spatial, 43
relationism, 18, 28, 41, 196

and possibilities, 32
defined, 27
negative message of, 32
spatial, 28
relations
   external, 31, 40
   realist, 35
   spacetime, 29
   spatial, 28, 31, 45
      thing-to-thing, 28
   spatiotemporal, 46
   thing-to-space, 24
Resnick, R., 121
Ricci (Ricci-Curbastro, G.), 149
Riemann tensor, 162
Riemann, B., 36, 149
Riemannian space, 169
Riemannian spaces, 12
Riemannian spacetimes, 144
rigid motion, 175
Rindler acceleration, 85
Rindler motion, 95
Rindler, W., 54, 96, 114, 121
Rovelli, C., 156
Russell, B., 20, 38
Rynasiewicz, R., 146

Schild, A., 134
Schilpp, P. A., 6
Schrödinger, E., 138
Schutz, B., 96, 149, 158, 159
Schwarzschild, K., 24
Sen, A., 56
shift
   dynamic, 13
   kinematic, 13
shift arguments, 22
   unnatural longevity, 19
Shwarszchild spacetime, 127
Sklar, L., 30, 33
slowing of a moving clock, 192
snowstorm, 110, 111
Socrates, 100
Sorensen, R., 191
space

absolute, 6, 12
curved, 131
Euclidean, 15, 19, 21, 124, 171
infinite, 22
homogeneous, 22
metric, 36
Newtonian, 63
unthinkable, 6
variable curvature, 21
Spacetime
   Role in ontology, 196
spacetime, 1, 5, 8, 34, 44, 48, 56,
      58, 63, 94, 100, 101, 105,
      110, 157
   basic properties, 187
   immaterial, 43
   no ghost, 193
   non-dynamic, 4
   not hidden, 130, 133–134
   ontic type, 7
   ontologically unique kind, 2
   pseudo space, 67
   real, 1
   reality of, 71
   role in ontology, 2
   unique ontic type, 7
   unthinkable, 6
   variably curved, 24
spacetime curvature, 132
spacetime interval, 34
spacetime realism, 169
spacetime realists, 25
spacetime vector, 64
Special Relativity, 75, 78
Special Theory of Relativity, 4
Standard Model, 2
Stein, H., 56
structuralism, 193
substance, 40
substantivalism, 12, 18, 24, 28, 35,
      40, 42, 165, 193
   defined, 27
substrate, 40
symmetries
   thing-to-thing, 24

symmetry, 20, 21
    defined, 13
    global, 30
    Minkowski/Ignatowski, 69
    translation, 49

Taylor, E. F., 115, 121
temporal length, 116
tenseless speech, 110
Thomas precession, 56
Thorne, K., 96
tidal forces, 129
time
    A-theory, 114, 115
    absolute, 6
    B-theory, 9, 107, 111
    flow, 9
Tooley, M., 100, 102, 107, 110, 111, 117, 126
topology, 37
    Euclidean, 171
    global, 44
Torretti, R., 56, 142, 162
transformations
    active, 174
    passive, 174
trigonometry
    hyperbolic, 66
Twins Paradox, 5, 113

vectors
    addition of, 68, 70
    energy momentum, 63
    velocity, 63
velocity
    system, 68–71

Wald, R., 149, 177
Walter, S., 49
Weyl, H., 29
Wheeler J. A., 115
Wheeler, J. A., 5, 96, 121, 140
Whittaker, E. T., 49
Wigner rotation, 56
worldline, 63, 68, 91, 96, 131
    geodesic, 130

About the author

Graham Nerlich is Emeritus Professor of Philosophy in the University of Adelaide where he took his first degree. After postgraduate studies at Oxford he held several academic positions including the chair of philosophy at the University of Sydney before returning to the Hughes Chair of Philosophy at Adelaide. He is the author of numerous papers in leading journals, of *The Shape of Space* CUP 1976, 2nd edn. 1994, *Values and Valuing: speculations on the ethical life of persons* OUP 1989; *What Spacetime Explains: metaphysical essays on space and time* CUP 1994. He is a Fellow of the Australian Academy of the Humanities and a Founder of the Minkowski Institute. He has three sons and is married to Margaret Rawlinson.

www.ingramcontent.com/pod-product-compliance
Lightning Source LLC
Chambersburg PA
CBHW060655100426
42734CB00047B/1800